编委会
Editorial Board

主　编 张　娴
副主编 许　珂、赵　晔
编　委
盛爱萍、徐晓燕、郭长升、王莉花、喻梦成、蔡　超、何萌萌、李思逊、严　甜、胡正鹏、苏鹤勇、姚凤君、李　阳、丁盼盼、袁　珍、李　信、胡　昀、佟晶晶、尹　超、吴林梅、周垠全、顾佳磊、陈　翀、陈聪颖、吴　磊、李　娟、董　璐
摄　影 张超、王　澄、李　阳
顾　问 姚亚祥、孙二平、刘　坚
技术指导 王剑峰
技术支持
陆跃明、曹澄亮、马春燕、吕　敏、苏晓晴、乔建英、王纪峰、龚　洁、张　倩、张燕汾、凌华峰、杨春梅、周　珏、张爱凤
感　谢
闵行区规划和自然资源局、上海南虹桥投资开发有限公司、华漕镇、新虹街道、七宝镇、虹桥镇、古美路街道、梅陇镇、莘庄镇、莘庄工业区、颛桥镇、马桥镇、江川路街道、吴泾镇、浦锦街道、浦江镇

心动上海，闵行正当年
规划建设二十年

THE RATE OF SHANGHAI
The Prime Time of Minhang

上海闵行规划设计研究院有限公司　编著

中国·上海

同济大学出版社
Tongji University Press

序

　　上海的地理中心位置上，闵行这把"金钥匙"熠熠生辉。在二十多年的快速蜕变中，从一个郊区工业城镇发展到面积达 373 平方公里、人口约 250 万人的大区域，见证了上海迈向卓越全球城市的深刻大变革。

　　天行健，君子自强不息。1997 年地铁 1 号线莘庄站通车，点亮城乡建设大舞台。无数闵行人的奋斗和智慧，是狠抓机遇的努力不懈，是实践探索的梦想实现。近 30 平方公里制造业基地全面提质增速，近千万平方米商务商办助力现代服务业发展，8000 万平方米住宅营造宜居生活，虹桥枢纽以及地铁 1、2、5、8、10、12 号线等一系列基础设施支撑现代都市高效运营，闵行正展露出现代化城市的繁荣风采。

　　新时代潮涌东方，心动春申。围绕长三角区域一体化、自贸试验区等重大国家战略，市、区两级 2035 总体规划明确下一个二十年发展蓝图，按照上海"五个中心"战略部署，闵行区正朝着建设成为上海全球城市重要的战略支撑区，品质卓越、生态宜居的现代化新城区昂首前进。空间战略突出南北两翼创新发展，北部建设虹桥国际开放枢纽，南部建设科创中心核心区。288 平方公里的建设空间亟待升级、蓄势待发，115 平方公里的生态空间绿色环绕。未来，品质宜居、绿色繁荣、智慧美好的闵行正大踏步地走来。

　　恰在少年奋进，当求风华正茂。蓬勃发展的闵行，不易之中盛满自豪，是记录者，更是时代的幸运儿，是上海都市交响乐的动人乐章，是辛勤劳动追求中国梦的坚强基石。

同济大学教授
上海市人民政府参事
上海市规划委员会专家
2020 年 11 月 26 日

Preface

Located in the geographic center of Shanghai, Minhang performs a vital role in the area. In the rapid transformation of more than 20 years, Minhang has developed from a suburban industrial town to a large area with an area of 373 square kilometers and a population of about 2.5 million. It has witnessed Shanghai's profound transformation into an outstanding global city.

As celestial bodies maintain vigor through movement, a gentleman should pursue constant self-improvement. In 1997, Xinzhuang Station of Metro Line 1 opened to traffic, lighting up the big stage of urban and rural construction. The struggle and wisdom of countless Minhang people are reflected in the unremitting efforts to seize opportunities and the realization of the dream of practical exploration. The manufacturing base of nearly 30 square kilometers has improved the overall quality and growth rate, nearly 10 million square meters of commercial offices have helped the development of modern service industry, 80 million square meters of residential buildings have created a livable life, and a series of infrastructures, including the Hongqiao Comprehensive Traffic Hub, Metro Lines 1, 2, 5, 10 and 12, have supported the efficient operation of modern cities. Minhang is showing the prosperous style of modern cities.

China has been modernizing, which motivates Shanghai to develop faster. Focusing on major national strategies, such as the integrated regional development of the Yangtze River Delta and the Pilot Free Trade Zone, the master plan 2035 clearly defines the development blueprint of both Shanghai and Minhang for the next 20 years. According to the strategic deployment of "Five Centers" in Shanghai, Minhang District is heading forward to build itself into an important strategic support area for Shanghai and a modern new town with excellent quality and ecological livability. The spatial strategy highlights the innovation and development of the north and south areas, with the construction of the Hongqiao Comprehensive Traffic Hub in the north, and the construction of the core area of Technology Innovation Center in the south. The construction space of 288 square kilometers needs to be upgraded and ready to go; 115 square kilometers of ecological space are surrounded by green. In the future, Minhang, which is livable in quality, green, prosperous, smart and beautiful, is coming in great strides.

Minhang is just in the youth, forging ahead to seek prosperity. The vigorous development of the district is full of pride in the hard work. It is the recorder and the lucky of the times. It is the pride of Shanghai and the strong cornerstone of the hard work in pursuing the Chinese Dream.

TANG Zilai

Professor at Tongji University
Counsellor of Shanghai Municipal People's Government
Expert at Shanghai Planning Commission
November 26, 2020

前言

闵行区规划设计研究院自 1996 年建院以来，参与了大量闵行区的规划研究工作，希望以自身视角将这二十多年的规划建设成就展现在广大市民面前，为闵行区今后的发展贡献自己的力量。

本书由闵行概览、产业发展、社会民生、生态休闲、乡村蜕变、地区风采六个篇章组成。其中，闵行概览介绍闵行自建区以来的总体概况、空间战略、规划脉络、闵行区总体规划。产业发展讲述闵行区的产业变迁、空间布局，并重点介绍产业园区、商务区、商业综合体、商业街区、文化创意街区、酒店等项目。社会民生介绍闵行区的居住社区、综合交通、市政设施、公共服务设施等的建设情况。生态休闲介绍闵行区的生态网络体系、河湖水系、公园体系、绿道体系、休闲活动等。乡村蜕变介绍闵行区乡村发展、农业生产、乡村生活、村落农宅、田园风貌、美丽乡村建设。地区风采主要介绍闵行区十四个街镇的基本建设情况及特色风貌。

本书收录的数据资料截至 2018 年 12 月，各类规划相关的图文资料由闵行区规划和自然资源局提供，摄影作品均为闵行区规划设计研究院自摄。

本书在编制过程中，得到闵行区有关部门和各街镇的大力支持及闵行区规划和自然资源局的具体指导，在此表示由衷感谢。

Forword

Since its establishment in 1996, Minhang District Planning and Design Institute has participated in a large number of planning and research work in Minhang District, hoping to show the achievements of planning and construction in the past 20 years in front of the general public from its own perspective and contribute its own strength to the future development of Minhang District.

This book consists of 6 chapters: Minhang overview, industrial development, social livelihood, ecological leisure, rural metamorphosis and regional elegance. Among them, Minhang overview introduces the general introduction, spatial strategic, planning context, comprehensive plan of Minhang. Industrial development describes the industrial changes and spatial layout of Minhang District, and focuses on industrial parks, business districts, commercial complex, commercial streets, cultural & creative blocks, hotels and other projects. Social livelihood introduces the construction of residential communities, comprehensive transportation, municipal facilities and public service facilities in Minhang District. Ecological leisure introduces Minhang District's ecological network system, water system, park system, greenway system, leisure activities, etc. The rural metamorphosis introduces the rural development, agricultural production, country life, village farmhouse, landscape in countrysides and beautiful village construction in Minhang District. The regional elegance mainly introduces the basic construction and characteristics of 14 streets and towns in Minhang District.

As of December 2018, the data and materials included in this book are provided by Minhang District Planning and Natural Resources Bureau. All the photos are taken by Minhang District Planning and Design Institute.

In the process of compilation, this book has been greatly supported by relevant departments and towns of Minhang District, as well as the specific guidance of Minhang District Planning and Natural Resources Bureau. We would like to express our sincere thanks.

目录 Contents

序	005
前言	007

1 第1章 闵行概览

总体概况	014
空间战略	015
规划脉络	016
闵行区总体规划	018

2 第2章 产业发展

工业变迁	026
商贸升级	028
产业空间布局	030
战略留白区	031
产业园区	032
商务区	035
商业综合体	046
商业街区	049
文化创意街区	051
酒店	053

3 第3章 社会民生

居住社区	058
综合交通	062
市政设施	074
公共服务设施	077

4 第4章 生态休闲

生态网络体系	096
河湖水系	098
公园体系	100
绿道体系	108
休闲活动	110

5 第5章 乡村蜕变

乡村发展	118
农业生产	120
乡村生活	122
村落农宅	123
田园风貌	125
美丽乡村	126

6 第6章 地区风采

虹桥商务区华漕镇	138
虹桥商务区新虹街道	142
七宝镇	146
虹桥镇	150
古美路街道	154
梅陇镇	158
莘庄镇	162
颛桥镇	166
莘庄工业区	170
马桥镇	174
江川路街道	178
吴泾镇	182
浦锦街道	186
浦江镇	190

后记	195
编者按	197

Preface	006
Forword	008

1 Part 1: Minhang Overview

General Introduction	014
Spatial Strategic	015
Planning Context	016
Comprehensive Plan of Minhang	018

2 Part 2: Industrial Development

Changes in Industry	026
Updates in Business	028
The Layout of Industrial Space	030
Strategic Blank Areas	031
Industrial Parks	032
Business Districts	035
Commercial Complexes	046
Commercial Streets	049
Cultural & Creative Blocks	051
Hotels	053

3 Part 3: Social Livelihood

Residential Communities	058
Comprehensive Transportation	062
Municipal Facilities	074
Public Service Facilities	077

4 Part 4: Ecological Leisure

Eco-network System	096
Water System	098
Park System	100
Greenway System	108
Leisure Activities	110

5 Part 5: Rural Metamorphosis

Rural Development	118
Agricultural Production	120
Country Life	122
Village Farmhouse	123
Landscape in Countryside	125
Beautiful Villages	126

6 Part 6: Regional Elegance

Hongqiao Central Business District of Minhang (Huacao Town)	138
Hongqiao Central Business District of Minhang (Xinhong Street)	142
Qibao Town	146
Hongqiao Town	150
Gumei Street	154
Meilong Town	158
Xinzhuang Town	162
Zhuanqiao Town	166
Xinzhuang Industrial Zone	170
Maqiao Town	174
Jiangchuan Street	178
Wujing Town	182
Pujin Street	186
Pujiang Town	190

Postscript	196
Editor's Notes	198

第1章
闵行概览

**PART 1
MINHANG
OVERVIEW**

总体概况
General Introduction

1992年，原闵行区和上海县合并，建立新的闵行区。辖区面积373.3平方公里，东与徐汇、浦东相接；南靠黄浦江与奉贤相望；西与松江、青浦接壤；北与长宁、嘉定毗邻。黄浦江纵贯南北，将闵行区划分为浦东、浦西两部分。

新闵行形似一把处于上海市域中心的"钥匙"，开启了现代化建设的新篇章。从1997年到2018年，闵行所辖行政区划从15个镇、3个街道，调整为9个镇、4个街道、1个市级工业区；总人口从57.4万人到254.4万人，增长近4.5倍；经济总量从136亿元到2014亿元，增长近15倍。

作为上海市唯一的国家产城融合示范区，今日的闵行综合交通发展凸显都市脉动，商圈林立展现时尚风貌，古镇新颜描绘历史长卷，蓝绿交织演绎生态家园。明天的闵行将建设成为充满活力、高效便捷、环境优美、城乡融合、社会和谐的生态宜居现代化新城区。

In 1992, a new Minhang District was established by merging the former Minhang District with Shanghai County. The district covers 373.3 square kilometers, surrounded by Xuhui and Pudong on the east, Fengxian on the south, Songjiang and Qingpu on the west, and Changning and Jiading on the north. The whole area is divided into Pudong and Puxi by Huangpu River, flowing from north to south and shifting westward between Minhang and Fengxian.

The outline of the new Minhang is like a "key", locating in the central position of Shanghai and becoming a pioneer in modernization construction. From 1997 to 2018, the area under its jurisdiction changed from 15 towns and 3 streets to 9 towns, 4 streets and 1 municipal-level industrial zone. Its population expanded from 0.574 million to 2.544 million, nearly an increase of 4.5 times, and the GDP of the district saw a dramatic rise from 13.6 billion to 201.4 billion, about 15-fold growth.

As the only National City-Industry Integration Demonstration Zone in Shanghai, Minhang nowadays is a region with vibrancy, fashion, historical connotations, and ecological environment by developing comprehensive transportation, building a forest of business districts, redesigning its ancient town, creating green space and reducing air pollution. Minhang will be developed into a modern habitable green district with vitality, high efficiency and convenience, beautiful environment, achieve success in urban-rural integration and construct a harmonious society.

莘庄立交
Xinzhuang Interchange

区位图
Site Location

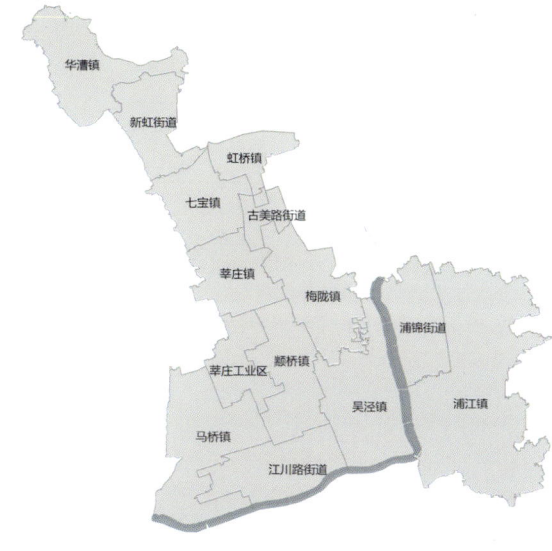

行政区划图
Administrative Division

空间战略
Spatial Strategic

　　闵行近 20 年的空间发展，是上海全球城市繁荣繁华的一个缩影。从卫星城到郊区新城，从西南一隅的工业小镇到南北两片创新大发展，城市空间急剧变迁，产业功能全面升级，彰显着经济、社会、文化的时代跨越。

　　顾往看今，城市空间格局主要 3 个阶段：

　　初期的以产带城，新中国成立初期，上海以工业城市为定位，闵行、吴泾卫星城规划的批复，明确了闵行以工业带动城市的发展战略。在此背景下，闵行和吴泾两大工业基地构建起闵行南部城市发展骨架，带动闵行南部的发展。

　　跨世纪的以产兴城，园区经济高速增长推动快速城市化。闵行开发区、莘庄工业区、紫竹高新区、漕河泾浦江科技园等带领区域经济再创新高。近 20 年间，工业用地增量超过 70 平方公里，国内生产总值（GDP）增长近 10 倍。城市功能快速扩张，莘庄、七宝等地人口快速增长，成为上海中心城区人口疏解和外来人口导入的主要承载地之一，逐渐向产城融合的郊区新城迈进。

　　21 世纪的产城融合，2010 年至今，走上城市空间与产业整合发展的道路。是镇域经济到区域经济的转变重点，城市功能进一步整合，形成了北部大虹桥现代服务功能区、南部大紫竹科技创新功能区、东部大浦江城乡统筹功能区三大空间板块。国家级产城融合示范区获批，助力形成城市功能更加完善、公共服务更加完备、城市生活更加美好的现代化主城区。

　　面向未来，上海全面建设卓越的全球城市，闵行着力打造品质卓越、生态宜居的主城区，开启城市发展新篇章。这里是闵行，是历史，是文化，也是 254 万人的美好向往；这里是品质，是绿色盈盈，是交通发达，也是公共服务的人人共享；这里是生活，是明天，是希望，也是最可期待的理想之地。

The spatial development of Minhang during the past 20 years is a microcosm of the prosperity and prosperity of the global city of Shanghai. From a satellite city to the suburban new city, it has experienced the innovative development, developing from an industrial town in the southwest to an area extending from the north to the south, the urban space has changed dramatically, and industrial functions have been fully upgraded, highlighting the changes in economy, society and cultures.

The change of urban spatial pattern goes through three stages:

Developing the area based on industry, in the early days of the founding of the People's Republic of China, Shanghai was positioned as an industrial city, and the approval of the plans of setting Minhang and Wujing Satellite City clarified the development strategy of Minhang's industrial-driven city. In this context, the two major industrial bases of Minhang and Wujing build the framework of urban development in the south of Minhang and drive the development of southern Minhang.

The city is booming with industry, and the rapid economic growth of the park promotes rapid urbanization. Minhang Development Zone, Xinzhuang Industrial Zone, Zizhu High-tech Zone, Caohejing Pujiang High-tech Park, etc. have led the regional economy to new heights. In the past two decades, the increase of industrial land has exceeded 70 square kilometers, and the gross domestic product (GDP) has increased by nearly 10 times. The rapid expansion of urban functions, the rapid population growth in Xinzhuang, Qibao and other places has become one of the main place for population dispersal and the introduction of foreign population in Shanghai's central urban area, and it is gradually moving towards a new suburban city, becoming a city-integration city.

Industry-city integration in the 21st century: since 2010, the area has embarked on the road of integrated development of urban space and industry. It is the focus of the transition from the township economy to the regional economy, and the city functions are further integrated, forming three major space sections: the Dahongqiao modern service function zone in the north, the science and technology innovation function zone in the south, and the Dapujiang urban-rural integrated function zone in the east. The national-level demonstration zone for the integration of industry and city was approved to help modernize the city's functions, improve public services, and improve the city's life.

Facing the future, Shanghai is building an excellent global city in an all-round way. Minhang strives to create a main urban area with excellent quality and ecological livability, and opens a new chapter in urban development. Minhang is an area of history, culture, and a longing place for 2.54 million people; here one can enjoy quality life, livable environment, well-developed transportation, and public service; it is an ideal place where one can achieve his dreams and have a bright future.

规划脉络
Planning Context

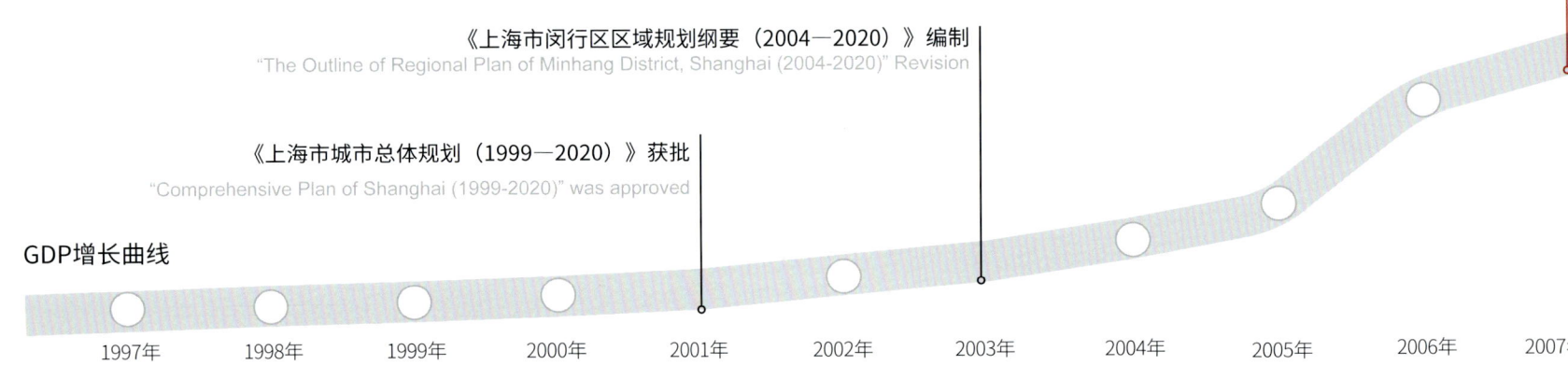

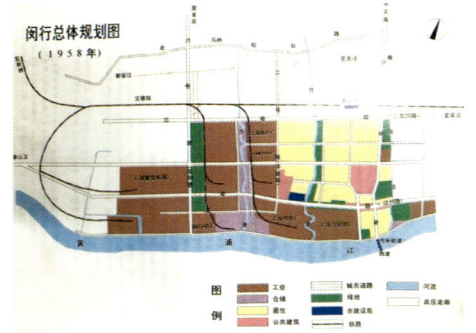

1958年闵行总体规划图
Comprehensive Plan of Minhang, 1958

1989年闵行总体规划图
Comprehensive Plan of Minhang, 1989

1999年上海市总体规划图
Comprehensive Plan of Shanghai, 1999

1999年上海市中心城总体规划图
Comprehensive Plan of Central Shanghai, 1999

《上海市城市总体规划（2017—2035）》获批
"Comprehensive Plan of Shanghai (2017-2035)" was approved

《上海市闵行区总体规划暨土地利用总体规划（2017—2035）》获批
"Comprehensive Plan and General Land-use Plan of Minhang District, Shanghai (2017-2035)" was approved

《虹桥商务区规划》获批
"The Plan of Hongqiao Business District" was approved

《上海市闵行区新城总体规划（2007—2020）》获批
"Comprehensive Plan of New City of Minhang District, Shanghai (2007-2020)" was approved

2008年　2009年　2010年　2011年　2012年　2013年　2014年　2015年　2016年　2017年　2018年　2019年

第1章　闵行概览　Part 1 Minhang Overview

闵行区总体规划
Comprehensive Plan of Minhang

新闵行成立以来，编制过两版经市政府批准的闵行区总体规划，即《闵行区区域总体规划实施方案（2006—2020年）》（简称"闵行2020"）以及《闵行区总体规划暨土地利用总体规划（2017—2035年）》（简称"闵行2035"）。

两版总规，充分体现了不同时代的发展要求。功能定位上，从具有辅城功能的现代化新城到全球城市的主城区，实现从追赶到超越的目标转变；发展思路上，从增量发展到存量更新、底线管控；编制内容上，从城市规划到规土合一，突显规划的公众性和协调性。

Two versions of comprehensive planning of Minhang District have been authorized by Shanghai Municipal Government since the establishment of the new district.

The two versions reflect different requirements under different historical backgrounds. From the perspective of functional orientation, Minhang has achieved remarkable achievements in developing the area from a modern auxiliary district in Shanghai into an international central downtown of global city. As to the ideas of development, Minhang changes its incremental development pattern by updating stocks and strictly controlling its baseline. Also, the district shifts its focus on growth management from urban planning to a new pattern—integration of urban planning and land use plan, thereby making Minhang a coordinated and inclusive district.

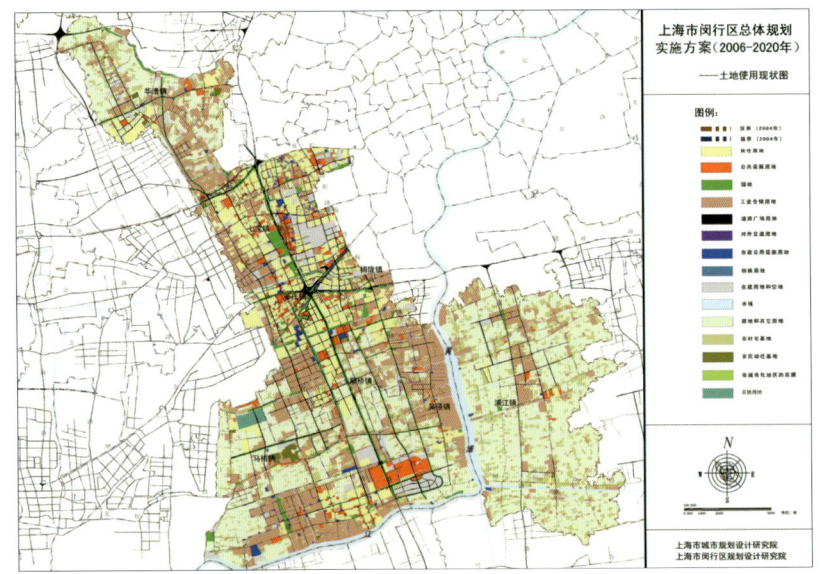

土地使用现状图（来源：闵行2020）
Land Use Status (From: Minhang 2020)

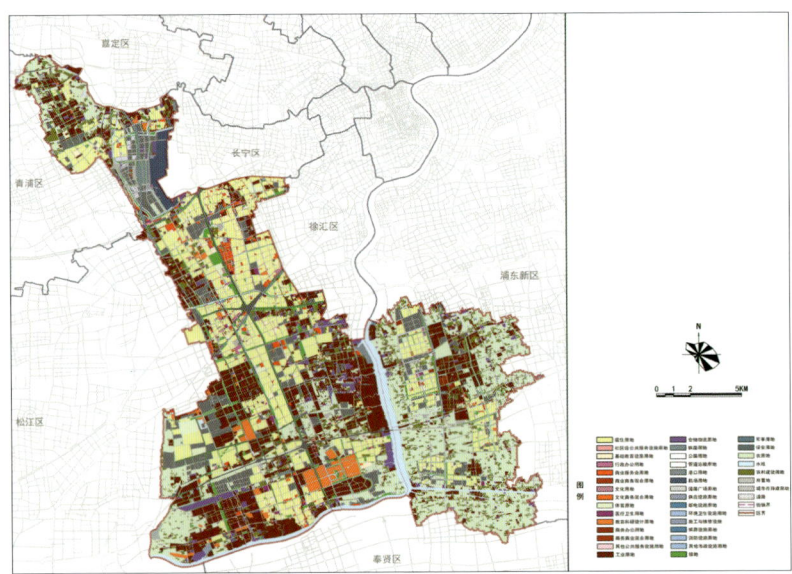

土地使用现状图（来源：闵行2035）
Land Use Status (From: Minhang 2035)

《闵行区区域总体规划实施方案（2006—2020年）》（简称"闵行2020"）
The Implementation Plan of the Comprehensive Plan of Minhang District, Shanghai, 2006–2020 (abbreviation: Minhang 2020)

规划目标：打造"航天闵行"，把闵行建成自主创新推动、产业集群发展、生态环境良好的具有新型辅城功能的现代化新城区。

现状规模：城市建设用地 183.7 平方公里。（2005 年）

规划规模：至 2020 年常住人口 200 万人，城市建设总用地 277 平方公里。

The object of the planning: build "Aerospace Minhang" and develop Minhang into a modern auxiliary area with capacity for independent innovation, industrial cluster's development, and habitable ecological environment.

Scale of current status: 183.7 square kilometers urban land for construction in 2005.

Scale in plan: 277 square kilometers land in total for urban construction, and 2 million permanent resident population until 2020.

《闵行区总体规划暨土地利用总体规划（2017—2035年）》（简称"闵行2035"）
Comprehensive Plan and General Land-use Plan of Minhang District, Shanghai, 2017–2035 (abbreviation: Minhang 2035)

规划目标：把闵行建设成为上海卓越的全球城市、具有世界影响力的社会主义现代化国际大都市的重要战略支撑区，品质卓越、生态宜居的现代化新城区。

现状规模：城市建设用地 257.9 平方公里。（2013 年）

规划规模：至 2035 年常住人口 254 万人，城市建设总用地 288.8 平方公里。

The object of the planning: build Minhang into an important strategic area of support for an outstanding global city, a modern socialist cosmopolis with world influence, and a new modern urban area with excellent quality and ecological livability.

Scale of current status: 257.9 square kilometers urban land for construction in 2013.

Scale in plan: 288.8 square kilometers land in total for urban construction, and 2.54 million permanent resident population until 2035.

土地使用规划图（来源：闵行2020）
Land Use Planning (From: Minhang 2020)

土地使用规划图（来源：闵行2035）
Land Use Planning (From: Minhang 2035)

锦江乐园
Jinjiang Park

蒲汇塘桥
Puhuitang Bridge

闵行中学
Minhang High School

闵行卫星城
Minhang Satellite Town

上海南方商城
Shanghai Nanfang Shopping Mall

地铁1号线延伸段试通车典礼
The Inaugural Ceremony of Extended Section of Metro Line 1 in Test Phase

闵行经济技术开发区
Minhang Economic & Technological Development Zone

莘庄立交
Xinzhuang Interchange

上海自行车赛场
Shanghai Cycling Stadium

第1章 闵行概览 Part 1 Minhang Overview

第 2 章
产业发展

**PART 2
INDUSTRIAL
DEVELOPMENT**

工业变迁
Changes in Industry

从传统制造到高端智造
From Traditional Manufacturing to High-end Intelligent Manufacturing

20世纪，闵行以"四大金刚"——电机厂、汽轮机厂、锅炉厂、重型机器厂为工业基础，大力发展传统制造业。21世纪以来，"航天闵行"的产业特色逐渐形成。十多个航天研究所及四大船舶研究所等军工企业和科研机构，在扩大研发和生产时纷纷搬至闵行。近年来，众多高新技术产业更是纷至沓来，重点聚焦发展高端装备、人工智能、新一代信息技术和生物医药四大主导产业。从"四大金刚"到闵行开发区、莘庄工业区，再到紫竹高新区、漕河泾浦江高科技园，这是一条从"制造"到"智造"的发展之路，为闵行的崛起和腾飞留下了浓墨重彩的一笔。

In the 20th century, Minhang strived to develop the traditional manufacturing industry based on four main factories – Electric Motor Manufacturing Works, Turbine Works, Boiler Works, and Heavy Machinery Factory, and since the 21st century, the industrial characteristics of "Aerospace Minhang" have gradually taken shape; military industrial enterprises and scientific research institutions, including more than a dozen aerospace research institutes and ship research institutes, moved to Minhang one after another when expanding researches and production. In recent years, many high-tech industries have come to Shanghai, focusing on the development of four leading industries of high-end equipment, artificial intelligence, new generation of information technology and biomedicine. Minhang District has succeeded in conversing its traditional manufacturing based on four factories into intelligent manufacturing with Minhang Development Zone, Xinzhuang Industrial Zone, Zizhu High-tech Zone, and Caohejing Pujiang High-tech Park, which have fostered the development of Minhang District.

上海锅炉厂老照片
Old Photo of Shanghai Boiler Works

吴泾工业区老照片
Old Photo of Wujing Industrial Zone

漕河泾浦江高科技园
Caohejing Pujiang High-tech Park

闵行开发区老照片
Old Photo of Minhang Development Zone

闵行工业区老照片
Old Photo of Minhang Industrial Zone

第2章 产业发展　Part 2 Industrial Development

商贸升级
Updates in Business

从"去上海"到"来闵行"
From "Go to Shanghai" to "Come to Minhang"

20世纪90年代初,主要商业设施为老闵行的"一号路"(江川路)、莘庄的海星商场等。就连看电影、吃饭、买衣服这样的小事,闵行人也要"去上海"——到徐家汇、人民广场或淮海路等市中心购物消费。21世纪以来,随着南方商城、七宝巴黎春天、莘庄仲盛商城等相继落成,闵行居民去市中心消费的比例逐渐下降。近年来,随着经济发展,涌现了大量商务商业集聚区,各类商业服务设施层出不穷,闵行人大部分的日常消费都能在闵行区内解决。同时,闵行也逐渐成为辐射城市外围的重要节点,实现了从"去上海"到"来闵行"的转变。

In the early 1990s, the main commercial facilities were Jiangchuan Road (the "No. 1 Road" of Minhang in the past), Xinzhuang Haixing Shopping Mall, etc., and local residents in Minhang who wanted to watch movies, eat, and buy clothes had to go to downtown areas of Shanghai, such as Xujiahui, People's Square, or Huaihai Road. Since the 21st century, the proportion of locals shopping to the city center gradually dropped with the building of Nanfang Shopping Mall, Qibao Paris Spring Shopping Center, Xinzhuang Zhongsheng Sky Mall, etc. In recent years, with the economic development, a large number of business and commercial areas have emerged, and the emerge of various types of commercial service facilities in the district have met most of the daily consumption of Minhang people. At the same time, Minhang has gradually become an important central area that attracts people to do shopping in Minhang instead of to the downtown area of Shanghai.

一号路老照片
Old Photo of No.1 Road

海星商场老照片
Old Photo of Haixing Shopping Mall

一号路现状照片
Status Photo of No.1 Road

海星商场现状照片
Status Photo of Haixing Shopping Mall

七宝万科广场
Qibao Vanke Shopping Mall

产业空间布局
The Layout of Industrial Space

产业园区布局
The Layout of Industrial Parks

形成一大产业基地、七大产业社区的产业布局体系。预期至 2035 年，全区工业用地规划总量为 33 平方公里（不含留白区域），占全区总建设用地的 11%。

Minhang District has developed an industrial layout system with an industrial base and seven industrial communities. It is expected that by 2035, the total area of planned industrial land in the region will be 33 square kilometers (excluding the blank area), accounting for 11% of the total construction land in the region.

商业商务集聚区布局
The Layout of Business and Commercial Areas

引导商务集聚，突出商务特色，形成 14 个高品质集中商务区。预期至 2035 年，全区商业设施总量为 1200 万平方米，商务设施总量为 1800 万平方米。

The district attracts business companies by highlighting its business functions, forming 14 high-quality centralized business districts. It is expected that by 2035, the total commercial facilities in the area will be 12 million square meters and the total business facilities will be 18 million square meters.

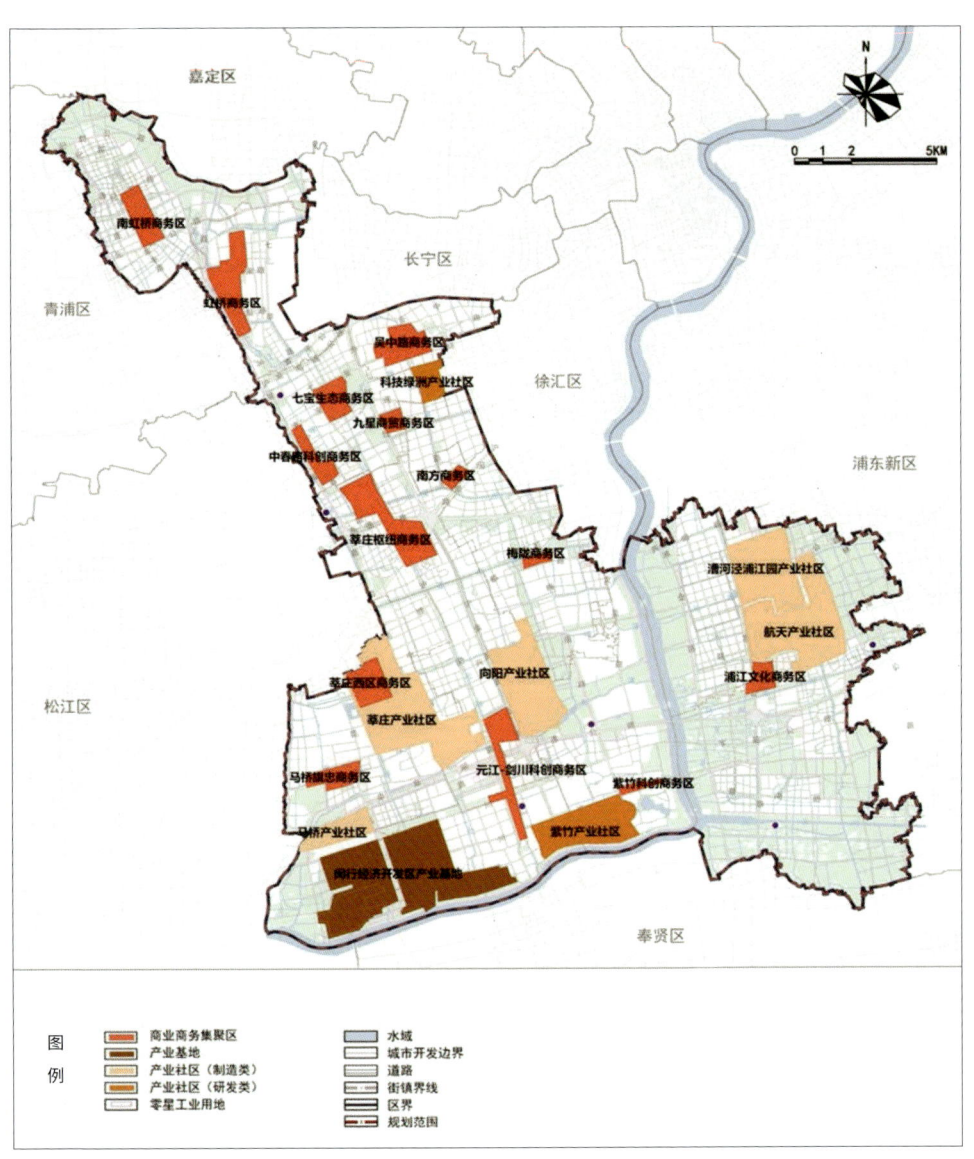

产业空间布局规划图（来源：闵行2035）
Industrial Spatial Planning (From: Minhang 2035)

战略留白区
Strategic Blank Area

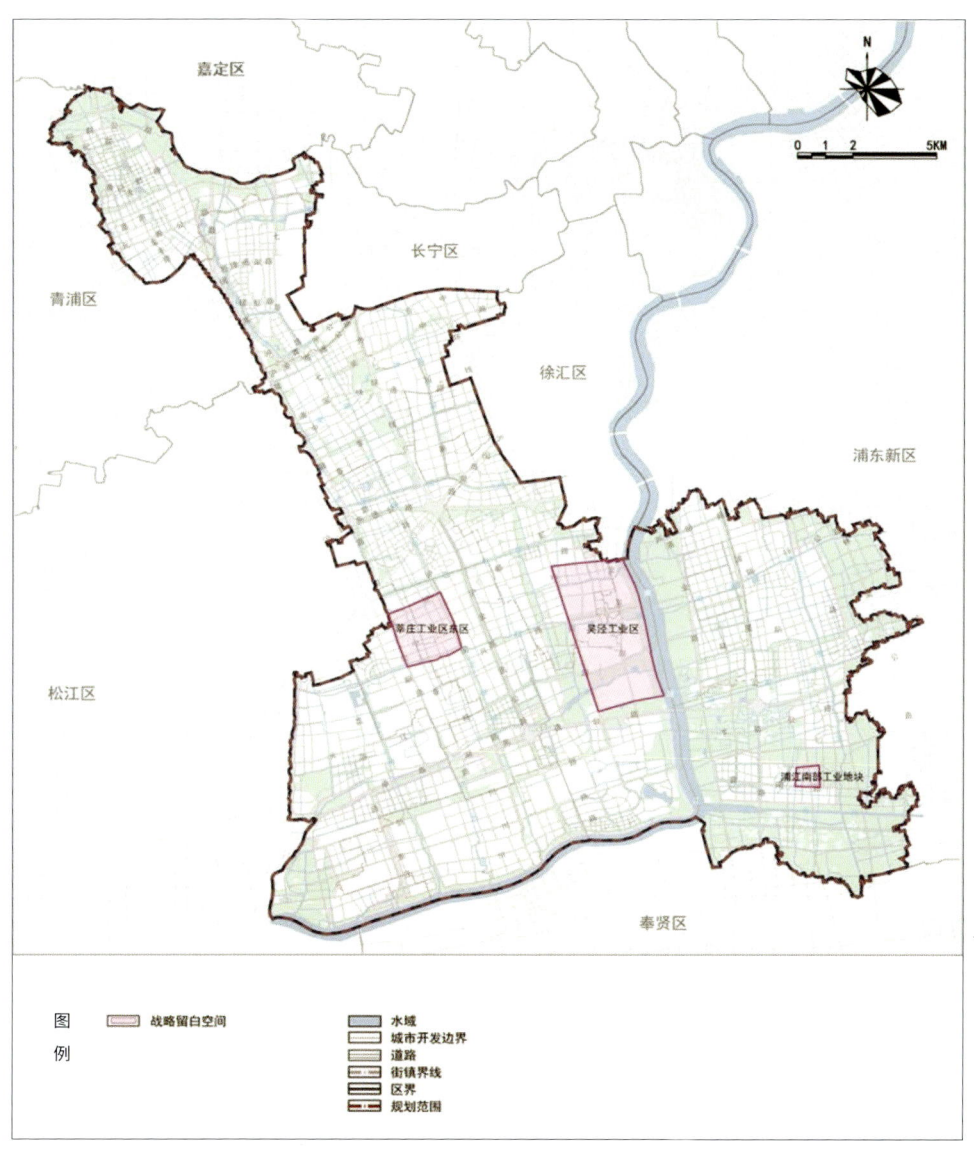

战略留白空间规划图（来源：闵行2035）
Strategic Blank Area Planning (From: Minhang 2035)

　　结合全区的功能布局调整，划定19.6平方公里建设用地作为战略留白区，包括吴泾工业区、莘庄工业区东区、浦江南部工业地块。其中，吴泾工业区逐步引导现有污染型工业用地退出，严格控制留白地区的建设活动，预留大型文体设施空间，作为大型综合性国际赛事场馆选址，实施生态修复。莘庄工业区东区和浦江南部工业地块预留未来承担全区产业转型升级的重要空间。

In combination with the adjustment of the functional layout of the entire area, 19.6 square kilometers of construction land was designated as areas to develop in the future, including Wujing Industrial Zone, the east zone of Xinzhuang Industrial Zone, and industrial land in southern Pujiang. Among them, the Wujing Industrial Zone gradually takes the lead in removing all polluting industry out of the area, strictly controls the construction on land to develop, reserves space for large-scale cultural and sports facilities as sites for large-scale comprehensive international competitions, and uses ecological remediation methods for reducing pollution in the district. The east zone of Xinzhuang Industrial Zone and industrial land in southern Pujiang are expected to become important spaces serving for industrial transformation and upgrading of the district.

产业园区
Industrial Parks

闵行经济技术开发区
Minhang Economic & Technological Development Zone

闵行经济技术开发区于 1986 年经国务院批准成为国家级经济技术开发区。园区现总面积约 3.5 平方公里，2006 年在临港成功扩区，规划面积为 13.3 平方公里。已基本形成以机电产业为主导、医药医疗产业和食品轻工产业为补充的三大产业板块，吸引了强生、西门子、米其林等世界 500 强企业入驻。

Minhang Economic & Technological Development Zone was approved by the State Council to become a national economic and technological development zone in 1986. The zone now has a total area of about 3.5 square kilometers, and in 2006, it was successfully expanded in Lingang area with a planned area of 13.3 square kilometers. The development zone has now basically formed three major industry sectors dominated by the electromechanical industry and supplemented by the pharmaceutical industry, medical industry, and food industry, which have attracted companies of Fortune Global 500, including Johnson & Johnson, Siemens, Michelin, etc.

闵行经济技术开发区鸟瞰图
Aerial View of Minhang Economic & Technological Development Zone

漕河泾浦江高科技园
Caohejing Pujiang High-tech Park

漕河泾浦江高科技园 2004 年经国务院批准成为国家级开发区。园区面积约 10.7 平方公里，现已基本形成以电子信息产业为依托，生物医药、节能环保、新能源、汽车研发为主体，现代生产性服务业为配套的产业格局，吸引了包括普洛斯、英业达、阿尔斯通等众多知名企业入驻。

Caohejing Pujiang High-tech Park was approved by the State Council to become a national development zone in 2004. The high-tech park covers an area of about 10.7 square kilometers, and now by giving priority to biomedicine, energy conservation and environment protection, new energy, and automobile research and development, it has basically formed an industrial structure based on the electronic information industry and supported by modern producer services, attracting many well-known enterprises to settle in, including Global Logistics Properties, Inventec and Alstom, ect.

漕河泾浦江高科技园鸟瞰图
Aerial View of Caohejing Pujiang High-tech Park

紫竹高新技术产业开发区
Zizhu High-tech Industrial Development Zone

紫竹高新技术产业开发区于2006年经国务院批准成为国家级开发区。园区一期规划面积约13平方公里，重点围绕集成电路与软件、新能源、航空、数字内容、新材料和生命科学等六大主导产业，吸引了包括英特尔、微软、中国商飞等一大批国内外知名研发机构和企业入驻。

Zizhu High-tech Industrial Development Zone was approved by the State Council in 2006 to become a national development zone. The first phase of the development zone has a planned area of 13 square kilometers, focusing on six leading industries such as integrated circuits and software, new energy, aviation, digital content, new materials and life sciences. It has attracted a large number of well-known research and development institutions and enterprises both at home and abroad, such as Intel, Microsoft, and Commercial Aircraft Corporation of China.

紫竹高新技术产业开发区鸟瞰图
Aerial View of Zizhu High-tech Industrial Development Zone

莘庄工业区
Xinzhuang Industrial Zone

莘庄工业区于 1995 年经上海市政府批准成为市级工业区。园区总开发范围 22.9 平方公里，其中本部面积约 17.9 平方公里，已基本形成以信息产业、新型材料、机电及汽车配件为主导的产业格局，拥有平板显示产业基地和航天新区两大高地，吸引了包括华硕电脑、大金空调、广电集团等众多知名企业入驻。

Xinzhuang Industrial Zone was approved as a municipal level industrial zone by Shanghai Municipal Government in 1995. The total development area of the industrial zone is 22.9 square kilometers, of which the area of the headquarters is about 17.9 square kilometers. It has formed an industrial pattern dominated by the information industry, new materials, electromechanical and auto parts, with two advantages—an industrial base of FPD and a new area for aerospace. These benefits help Xinzhuang Industrial Zone attract many well-known companies including ASUS, DAIKIN, and SVA Group.

莘庄工业区鸟瞰图（来源：莘庄工业区管委会）
Aerial View of Xinzhuang Industrial Zone (From: Xinzhuang Industrial Zone Management Committee)

商务区
Business Districts

虹桥商务区
Hongqiao Business District

虹桥商务区涉及闵行、长宁、嘉定、青浦四个区，其核心区和南虹桥片区位于闵行区。依托虹桥综合交通枢纽、国家会展中心等重大功能性项目，目前已经成为总部办公、高端商务、现代服务业、科技创新产业汇聚的聚集区。未来将承担国际开放枢纽的功能，成为引领长三角一体化发展的龙头，打造高效绿色的国际交通枢纽区、开放引领的国际会展贸易区、创新共享的世界级商务区和生态宜居的主城片区。

The Hongqiao Business District involves Minhang, Changning, Jiading, and Qingpu. Its core area and Southern Hongqiao area are located in Minhang District. Relying on major functional projects such as the Hongqiao Comprehensive Traffic Hub and the National Convention and Exhibition Center, it has now become a gathering area for headquarters office, high-end business, modern service industry, and scientific and technological innovation industries. In the future, it will assume the function of an international open hub, becoming a leader in the integrated development of the Yangtze River Delta, creating an efficient and green international transportation hub area, an open and leading international convention and exhibition trade area, an innovative and shared world-class business area, and an ecologically livable central downtown.

虹桥商务区街景（来源：新虹街道办事处）
Street View of Hongqiao Business District (From: Xinhong Street Office)

整体鸟瞰效果图（来源：南虹桥地区城市设计）
Overall Aerial View Rendering (From: Urban Design of Southern Hongqiao Area)

虹桥商务区街景
Street View of Hongqiao Business District

南方商务区
Nanfang Business District

南方商务区位于地铁 1 号线莲花路地铁站北侧，以百联南方购物中心、友谊商城及莲花国际广场为核心，是闵行传统的地区商业中心。随着中庚漫游城、城开中心以及全长 1229 米的空中连廊和 8 万平方米城市公共绿地的建设，南方商务区将成为"商务及商业功能集聚的上海西南地区商业中心"。

Nanfang Business District, located north of Lianhua Road Subway Station of Shanghai Metro Line 1, is an established regional commercial center in Minhang, taking Bailian Nanfang Shopping Center, Friendship Shopping Mall and Lotus International Plaza as its core. With the construction of Zhonggeng Manyou City, U Center City, a 1229-meter Air Pedestrian Corridor, and 80,000 square meters of greenland, the Nanfang Business District will become a business center in southwest Shanghai with a gathering of business and commercial functions.

整体升级改造
Upgrading & Rebuilding

万源路扩建为双向四车道、中庚漫游城开幕
Wanyuan Road & Zhonggeng Manyou City

1999年，百联南方购物中心
Bailian Nanfang Shopping Center, 1999

2008年，友谊商城
Friendship Shopping Mall, 2008

九星商务区
Jiuxing Business District

　　九星商务区是集体经济发展转型的样板地区。九星从最早的农贸商行、综合贸易市场逐步发展为中国市场第一村，成长为上海三大建材市场之一。该区于2017年转型，将依托自身优势，围绕家居建材流通产业主题，建设"一站式国际家居贸易中心"。

　　Jiuxing Business District is a model area for the development and transformation of the collective economy. Jiuxing has gradually developed from the earliest agricultural trade firm and comprehensive trade market into the "first village" in the Chinese market, and has grown into one of the three major building materials markets in Shanghai. Relying on its own advantages, Jiuxing achieved success in transformation in 2017; it has built a "one-stop international home trade center" in home-building materials circulation industry.

鸟瞰效果图（来源：九星地区城市设计）
Overall Aerial View Rendering （From: Urban Design of Jiuxing District）

九星地区卫星影像变迁（来源：Google Earth）
Changes of Satellite Images in Jiuxing District (From: Google Earth)

用地规划图（来源：九星地区控规局部调整）
Land Use Planning (From: Control Detail Plan of Jiuxing District)

历经十余年时间，九星成为中国市场第一村
After More Than 10 Years, Jiuxing Has Become the First Village in the Chinese Market

2017年，九星市场开始拆除
Jiuxing Market Began to Dismantle in 2017

莘庄商务区
Xinzhuang Business District

莘庄商务区是莘庄主城副中心的重要组成部分，主要以发展现代服务业为主导，集商务办公、会议论坛、科技研发、总部基地、商业休闲、文化交流等多元功能为一体。园区以战斗湖为中心，采用"低密度、高绿化"的规划理念，着力构筑人、水、自然和谐共生的公共活动与商务休闲空间。

Xinzhuang Business District is an important part of the sub-center of downtown Xinzhuang. It is mainly dominated by the development of modern service industries, integrating multiple functions such as business office, conference forum, technology research & development, headquarters base, business leisure and cultural exchange, etc. The area takes the Zhandou Lake as its center, adopts the planning concept of "low density and high greening rate", and focuses on building public activities and business and leisure spaces where citizens, water and nature harmoniously coexist.

鸟瞰效果图（来源：莘庄商务中心详细规划设计）
Overall Aerial View Rendering (From: Detailed Planning & Design of Xinzhuang Business District)

莘庄商务区
Xinzhuang Business District

莘庄商务区
Xinzhuang Business District

第2章 产业发展　Part 2 Industrial Development

商业综合体
Commercial Complexes

　　闵行的新生代商业综合群体日益崛起，虹桥天地、仲盛、凯德龙之梦热度不减，宝龙城、七宝万科、新华联人潮涌动，维璟广场、城开中心、颛桥万达如火如荼，还有超大体量的爱琴海购物公园、万象城已成为炙手可热的网红打卡点，正在建造中的"天荟"更是蓄势待发。这些火爆的商业综合体为闵行的现代服务业注入了新的活力，也将成为生态宜居主城区的发展新动力。

　　Minhang's new-generation commercial complex is on the rise, the Hub of Hongqiao, Zhongsheng Sky Mall, Capita Shopping Mall have continued to thrive, the crowds of Powerlong Masion, Qibao Vanke, Xinhualian are surging, and Westlink Shopping Center, U Center City, Wanda Plaza (Zhuanqiao Branch) are in full swing. The super-large Aegean Shopping Park and the Mixc Shopping Center have become hot spots for online celebrities, and the "Tianhui" under construction is ready to go. These hot commercial complexes have injected new vitality into Minhang's modern service industry and will also become a new driving force for the development of ecologically livable main urban areas.

爱琴海购物公园
Aegean Shopping Park

万象城
The Mixc Shopping Center

凯德龙之梦购物中心
Capita Shopping Mall

商业街区
Commercial Streets

老外街——虹梅路休闲街
Foreign Street—Hongmei Leisure Street

地处虹桥镇，长约 500 米，采用欧陆风情建筑风格，融合国际多元文化，是展示各国餐饮美食文化及风俗民情的特色街道。

Foreign street is located in Hongqiao Town, with a length of about 500 meters. The architecture in Hongmei Leisure Street adopts European styles and integrates various cultural elements of countries both at home and abroad, making it a distinctive street showing the food cultures and customs of different countries.

虹梅路休闲街
Hongmei Leisure Street

阿拉城
Ala Town

地处虹桥镇，采用装饰艺术的建筑风格，以餐饮美食、文娱生活、体验零售等多种业态为主，通过"24 小时不间断营业"的经营方式，力图打造夜上海的特色消费示范区。

Ala Town is located in Hongqiao Town, adopting the art deco architectural style. It mainly focuses on a variety of industries such as catering industry, cultural life, recreation, experiential retail, etc., and strives to create a demonstration area with its characteristics in Shanghai with a business practice of "open 24/7".

阿拉城
Ala Town

韩国街——虹泉路
South Korea Street—Hongquan Road

地处虹桥镇，周围有3万—4万韩国籍居民居住，汇集了多个韩国社区，以韩国人经营的沿街商铺为主，具有浓厚的韩国风情。

Hongquan Road is located in Hongqiao Town, there are 30-40 thousand Korean residents living in the area surrounding the street, bringing together a number of Korean communities, and street shops are mainly run by Koreans, with a strong Korean flavor.

虹泉路
Hongquan Road

金平路
Jinping Road

地处江川路街道，采用地中海与英伦两种建筑风格，是一条集购物、餐饮、娱乐、休闲等功能于一体的商业街区。

Jinping Road is located in Jiangchuan Street, and it adopts two architectural styles, namely, the Mediterranean style and the British style, becoming a commercial block that integrates functions, such as shopping, dining, entertainment, and leisure.

金平路
Jinping Road

文化创意园区
Cultural & Creative Blocks

近年来，各类文创载体在闵行应运而生、集聚发展，涌现出光华路文创街区、麦可将文创园等一批具有特色的文化创意园区。它们汇聚了新兴的产业元素、蕴含着丰富的艺术文化，更提升了人们的生活品位。

In recent years, various cultural and creative carriers have emerged and gathered in Minhang, and a number of unique cultural and creative parks have emerged, such as Guanghua Road Cultural & Creative Block and MT389 Cultural & Creative Park, etc. These places bring together new industrial elements, rich artistic culture, and improve people's taste in life.

光华路文创街区
Guanghua Road Cultural & Creative Block

麦可将文创园
MT389 Cultural & Creative Park

酒店
Hotels

随着闵行经贸和旅游业的蓬勃发展，酒店行业应时而上，大量高品质的商务、度假酒店落地闵行，如养云安缦酒店、虹桥索菲特大酒店、诺宝中心酒店等，同时也产生了各具特色的农家乐、民宿等新业态。

With the vigorous development of Minhang's economy and tourism industry, the hotel industry has sprung up, and a large number of high-quality business and resort hotels have landed in Minhang, such as the Amanyangyun Hotel, the Sofitel Hotel, and the Noble Center Hotel, etc., and meanwhile farmhouses, homestays, etc. with unique features have also been created in the area.

养云安缦酒店
Amanyangyun Hotel

第2章　产业发展　Part 2　Industrial Development

索菲特酒店
Sofitel Hotel

诺宝中心酒店
Noble Center Hotel

第 3 章
社会民生

PART 3
SOCIAL
LIVELIHOOD

居住社区
Residential Communities

随快速城镇化为闵行带来了大量人口导入，全区人口从 1997 年的 57.4 万人增长至 2018 年的 254.4 万人，闵行已成为上海中心城人口疏解的主要承载区、新上海人的家园。

Rapid urbanization has brought a large amount of population to Minhang. The population of the entire district has increased from 0.574 million in 1997 to 2.544 million in 2018. The district has become the home of new Shanghainese, serving as the main area for population dispersal for the central area of Shanghai.

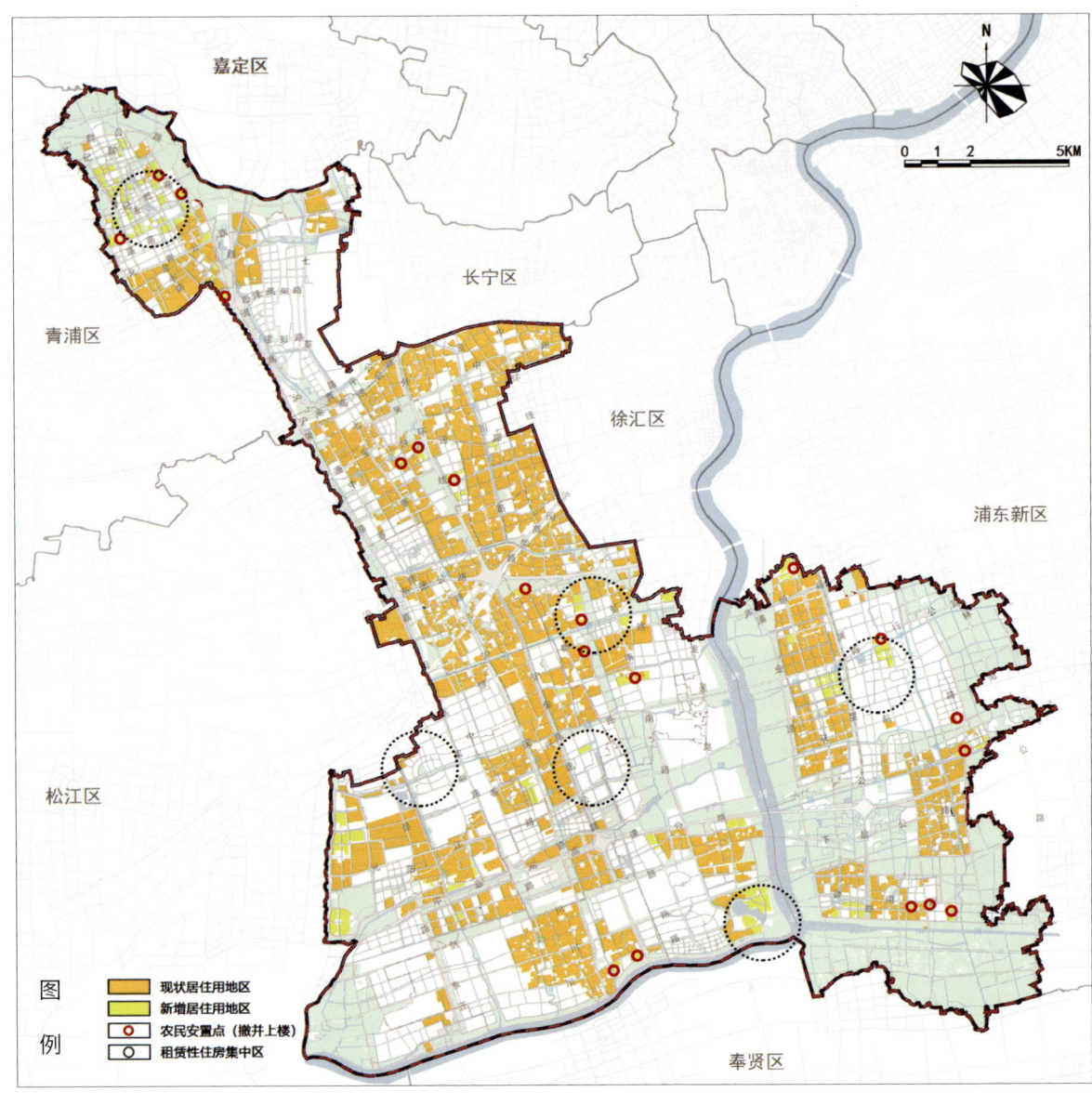

住房布局规划图（来源：闵行2035）
Residential Layout Planning (From: Minhang 2035)

全区已建住宅 8000 多万平方米,未来还将建设 1500 万平方米,涉及住宅用地共 79 平方公里。未来将进一步提升住宅品质,加强居住结构多样性,满足多元化需求,打造宜居的生活环境。

The district has built more than 80 million square meters of residential buildings, and will build 15 million square meters in the future, covering a total of 79 square kilometers of residential land. In the future, the quality of housing will be further improved, the diversity of living structures will be strengthened to meet diversified needs, and a livable living environment will be created.

颛桥镇住区
Residential Area of Zhuanqiao Town

保障性住房
State Housing

闵行区在探索和实践中完善了动迁安置房、经济适用房、廉租房、公共租赁房"四位一体"的保障性住房体系，为优化居住条件、改善民生不懈努力。2017年开始，结合轨道交通站点、产业园区、商务区等，布局建设社会租赁房，为人才引进提供更多空间。

Through exploration and practice, Minhang District has improved the "four-in-one" affordable housing system, which involves relocation housing, affordable housing, low-rent housing, and public rental housing, making unremitting efforts to optimize living conditions and improve people's livelihood. From 2017, in conjunction with rail transit stations, industrial parks, business districts, etc., layout and construction of social rental housing will be built to accommodate talents introduced.

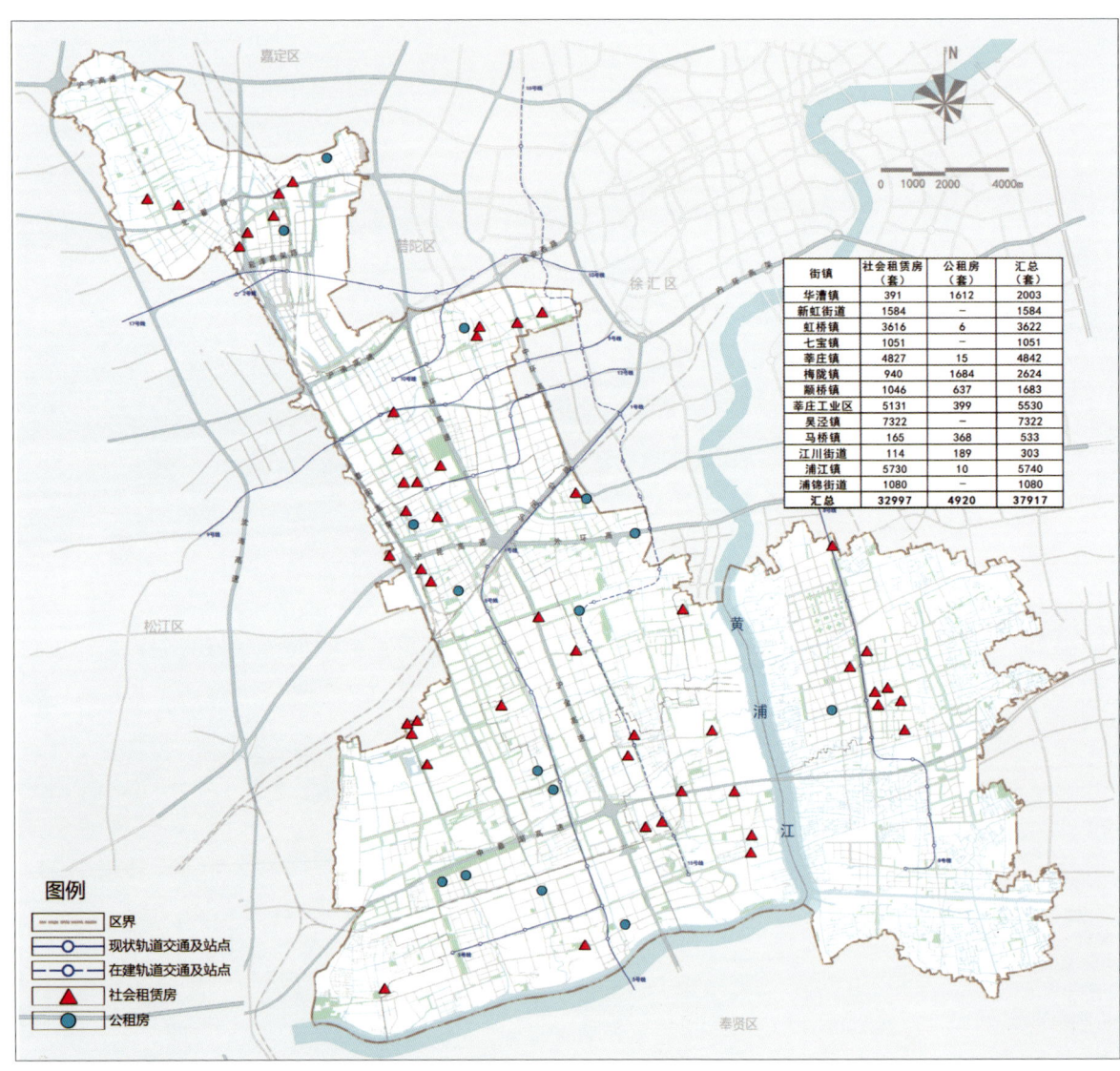

闵行区租赁住房地图（来源：上海市闵行区房管局）
Rental Housing Map of Minhang District (From: Shanghai Minhang District Housing Authority)

大型居住社区
Large Residential Communities

2009 年，上海市启动了大型居住社区选址与建设工作，10 年来在浦江、马桥、颛桥、梅陇等地落实了 9 处大型居住社区，构建了一个个功能完善、交通便捷、生态宜居、活力繁荣的城市社区，为中低收入及动迁人群提供了住房保障。

In the past 10 years since 2009 when Shanghai started the site selection and construction of large residential communities, 9 large residential communities have been settled in Pujiang, Maqiao, Zhuanqiao, Meilong and other places, forming a fully functional urban community with convenient transportation, ecologically livable environment, vibrancy and prosperity, which provides housing security for low-and-middle-income residents, as well as relocated people.

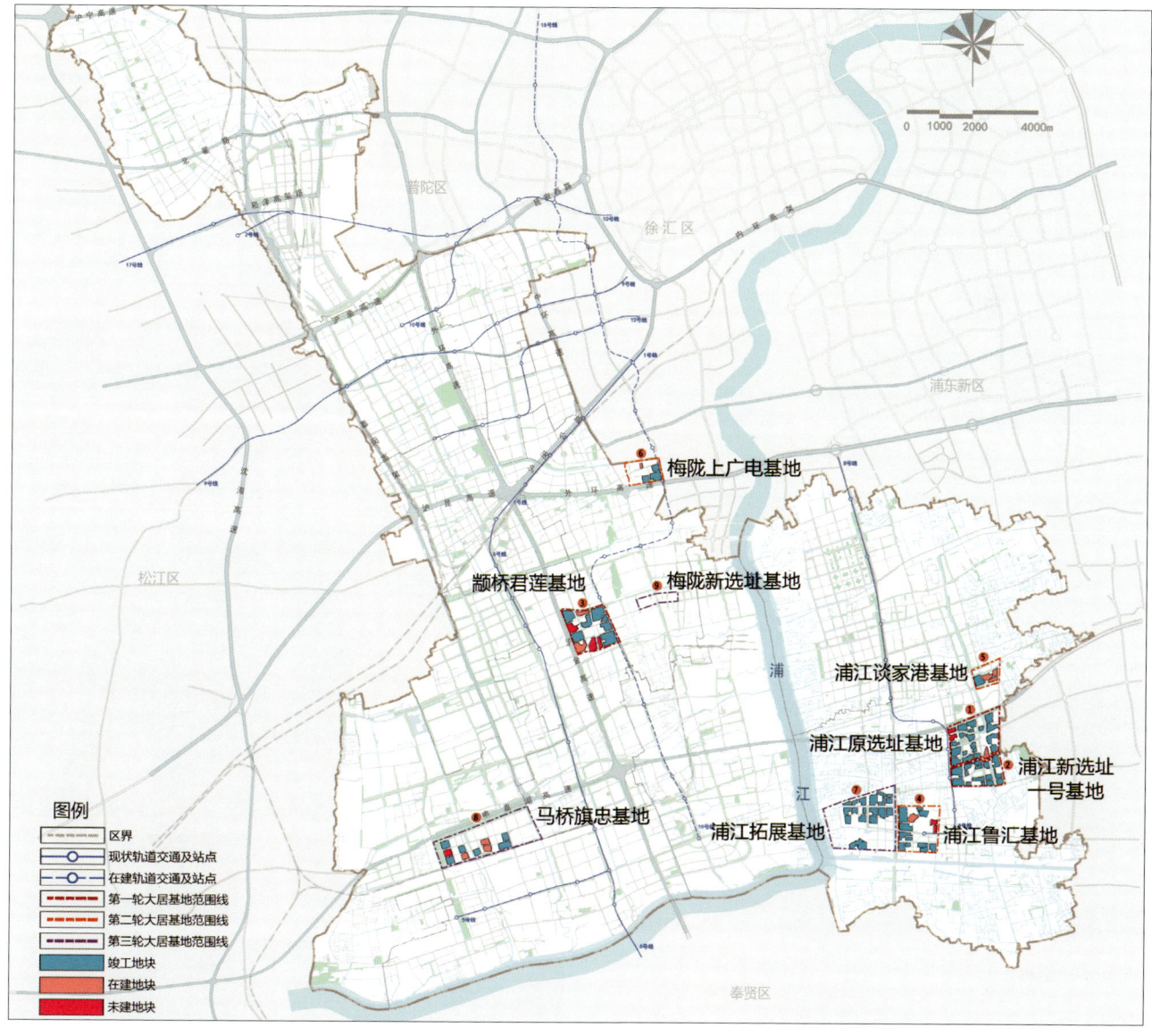

闵行区大型居住社区分布图（来源：来源：上海市闵行区房管局）
Distribution Map of Large Residential Communities in Minhang District (From: Shanghai Minhang District Housing Authority)

综合交通
Comprehensive Transportation

随着城市化进程的加快，闵行进入综合交通快速发展时期。2010年虹桥综合交通枢纽的建成，凸显了地区交通区位优势，提升了对外交通便利度。目前，以沪渝、沪昆为代表的高速公路和以嘉闵高架、虹梅南路高架为代表的快速路相继通车，标志着高、快速路网络的基本形成。全区轨道交通运营总长度已达63公里，轨道交通覆盖率稳步提升。未来，将构建以公共交通为主体，"集约低碳、集疏高效、内外畅达、接轨中心城"的一体化综合交通体系。

With the acceleration of urbanization, Minhang has entered a period of rapid development of comprehensive transportation. The completion of the Hongqiao Comprehensive Traffic Hub in 2010 highlights the advantages of regional transportation and improves the convenience of external transportation. At present, expressways represented by Shanghai-Chongqing and Shanghai-Kunming Expressways and fast roads represented by Jiamin Elevated Road and Hongmei Elevated Road have opened to traffic in succession, marking the formation of a transportation network with both expressways and fast roads. The total length of rail traffic in operation in the region has reached 63 kilometers, and the rail coverage has increased steadily. In the future, Minhang will construct an integrated and comprehensive transportation system with public transportation as the main body, which is low-carbon, efficient, accessible from all directions.

8号线轨道与列车
Track and Train of Metro Line 8

沪闵路、1号线远眺
A Panoramic View of Humin Road and Metro Line 1

道路交通
Road Traffic

目前,"高、快速路—主干路—次干路—支路"各级路网已基本成型,未来将通过整合挖潜、扩能提升、打通完善,形成结构基本合理、路网相对均衡的道路交通网络。规划道路网总规模 1500 公里以上,城市开发边界内全路网密度大于 8 公里 / 平方公里。

At present, the road network at all levels of "high and fast roads, main roads, secondary roads, branch roads" has been basically formed. In the future, it will be integrated, tapped, expanded, improved, and connected to form a road system with a basically rational structure and a relatively balanced road traffic network. The total length of the road network in plan is more than 1500 kilometers, and the density of the entire road network within the boundary of urban development is greater than 8 kilometers per square kilometer.

莘庄立交
Xinzhuang Interchange

沪闵高架路
Humin Elevated Road

嘉闵高架、华翔路
Jiamin Elevated Road, Huaxiang Road

轨道交通
Rail Traffic

自地铁 1 号线莘庄站通车以来，全区已运营 8 条轨道交通线路，42 座地铁站，创造出多个"第一"，如我国第一条采用高架轻轨制式的城市轨道线——地铁 5 号线，上海第一条全自动无人驾驶的城市轨道线——浦江线。未来，闵行将进一步优化轨道交通网络，构建由 23 条轨道交通、9 条中运量交通组成的轨道 交通系统。

Since the opening of Xinzhuang Station of Metro Line 1, the region has operated 8 rail lines and 42 subway stations, many of which are pioneering work in China, for example, Metro Line 5 which is the first elevated railway, and Pujiang Line which is a fully automated driverless rail line. In the future, Minhang will further optimize the rail traffic network and build a rail traffic system consisting of 23 rail lines and 9 lines with medium transportation capacity.

8号线浦航路站
Puhang Road Station, Metro Line 8

综合交通枢纽：虹桥综合交通枢纽
Comprehensive Traffic Hub: Hongqiao Comprehensive Traffic Hub

虹桥综合交通枢纽是一座集民用航空、高速铁路、城际铁路、轨道交通、中运量交通、常规公交等多种交通方式于一体的现代化大型综合交通枢纽。目前，虹桥综合交通枢纽高峰日客流量已突破140万人次，未来将依托虹桥商务区，建设面向全球、引领长三角地区更高质量一体化发展的国际开放枢纽。

The Hongqiao Comprehensive Traffic Hub is a modern large-scale integrated traffic hub that integrates a variety of transportation modes such as civil aviation, high-speed railway, intercity railway, rail transit, medium capacity transit, and conventional public transportation. At present, the daily passenger flow volume at peak time at the Hongqiao Comprehensive Traffic Hub has exceeded 1.4 million passengers. In the future, it will rely on the Hongqiao Business District to build an international hub that faces the world and leads the integrative development with higher quality in the Yangtze River Delta.

整体鸟瞰效果图（来源：虹桥综合交通枢纽地区控制性详细规划）
Overall Aerial View Rendering (From: Control Detailed Planning of Hongqiao Comprehensive Traffic Hub Area)

虹桥火车站内景
Interior View of Hongqiao Railway Station

2019年，虹桥火车站
Hongqiao Railway Station, 2019

综合交通枢纽：莘庄综合交通枢纽
Comprehensive Traffic Hub: Xinzhuang Comprehensive Traffic Hub

以1号线莘庄站为核心的莘庄综合交通枢纽，是闵行区快速城市化的起点，也是莘庄城市副中心的重要组成部分，周边集聚行政、文化、商办等重要功能。未来将着力打造集城际铁路、轨道交通、中运量交通、常规公交于一体的综合客运枢纽，进一步助力城市发展。

Xinzhuang Comprehensive Traffic Hub with Xinzhuang Station of Metro Line 1 as the core is the starting point of rapid urbanization in Minhang District, and also an important part of Xinzhuang, serving as the sub-center of the district. In the future, efforts will be made to build a comprehensive passenger transport hub that integrates intercity railways, rail transit, medium capacity transit, and conventional public transport, and further promote urban development.

整体鸟瞰效果图（来源：闵行区莘庄综合交通枢纽项目修建性详细规划）
Overall Aerial View Rendering (From: Construction Detailed Planning of Xinzhuang Comprehensive Traffic Hub Project in Minhang District)

2019年，建设中的莘庄站
Xinzhuang Station Under Construction, 2019

越江通道
Tunnel Passage

目前已形成"三桥一隧"的黄浦江越江通道，即闵浦大桥、闵浦二桥、奉浦大桥和虹梅南路隧道。未来，随着正在建设中的昆阳路越江大桥、银都路越江隧道相继通车，进一步加强黄浦江两岸交通联系。

At present, there are three bridges and one tunnel crossing the Huangpu River, namely Minpu Bridge, Minpu No.2 Bridge, Fengpu Bridge, Hongmei South Road Tunnel. In the future, as the Kunyang River-crossing Bridge, Yindu Road Tunnel under construction are being opened to traffic, the transportation links across the Huangpu River will be further strengthened.

闵浦大桥
Minpu Bridge

连接闵行浦西与浦东地区，是上海外环高速公路组成部分之一，2010年通车运营。线路全长3610米，宽44米，上层桥面为双向八车道的高速公路；下层为双向六车道的城市快速路。

Connecting the Puxi and Pudong areas of Minhang, it is one of the components of Shanghai Outer Ring Expressway and opened to traffic in 2010. The total length of the line is 3610 meters and the width is 44 meters. The upper deck is a two-way eight-lane motorway; the lower deck is a two-way six-lane urban expressway.

闵浦二桥
Minpu No.2 Bridge

连接闵行区与奉贤区，2010年通车运营。线路全长4893米，主桥全长437米。桥面上层为双向四车道城市快速路；下层为双线轻轨。
Connecting Minhang District with Fengxian District, it was opened to traffic in 2010. The whole traffic line is 4893 meters long and the main bridge is 437 meters long. The upper deck is a two-way four-lane urban expressway and the lower deck is a two-lane light rail.

昆阳路越江大桥
Kunyang River-crossing Bridge

又名闵浦三桥，建成后将连接闵行区与奉贤区。线路全长2300米，桥面上层为双向六车道；下层两侧供行人与非机动车通行。
Also known as Minpu No.3 Bridge, it will connect Minhang District and Fengxian District. The total length of the line is 2300 meters, and the upper deck of the bridge is two-way six lanes; both sides of the lower layer are for pedestrians and non-motor vehicles.

虹梅南路隧道
Hongmei South Road Tunnel

是连接闵行、奉贤两区，穿越黄浦江的重要交通设施。2015年年底通车，全长约5260米。其中，主线隧道长度约3390米，双向六车道。
It is an important traffic facility connecting Minhang District and Fengxian District and crossing Huangpu River. It was opened to traffic at the end of 2015, with a total length of about 5260 meters. Among them, the length of the main line tunnel is about 3390 meters, with two-way six lanes.

市政设施
Municipal Facilities

市政基础设施建设与经济建设同步推进，闵行基本建立了较完善的市政设施体系。经过多年的发展，全年供水量由 0.35 亿吨增长至 2.53 亿吨，售电量由 25.01 亿千瓦·时增长至 175.15 亿千瓦·时，天然气用量由 0.07 亿立方米增长至 6.64 亿立方米。未来将遵循合理布局、远近结合、适度超前、共建共享、优化配置的原则，构建完善的基础设施保障体系，让城市生活更加安全、便捷、舒适、高效。

Municipal infrastructure construction and economic construction have been advanced simultaneously, and a relatively comprehensive municipal facility system has basically been established. After years of development, the annual supply of water has increased from 35 million tons to 253 million tons, the amount of electricity sold has increased from 2.501 billion kW·h to 17.515 billion kW·h, and the amount of natural gas has increased from 7 million m³ to 664 million m³. In the future, the development of infrastructure will follow the principles of rational layout, combination of distance and nearness, moderate advancement, achieving shared growth through collaboration and optimizing allocation to build a comprehensive infrastructure security system in order to make urban life safer, more convenient, comfortable, and efficient.

吴泾电厂
Wujing Power Plant

浦锦垃圾中转站
Pujin Waste Transfer Station

2019年，闵行自来水厂
Minhang Water Plant, 2019

2019年，上海广播卫星地球站
Shanghai Radio Satellite Earth Station, 2019

公共服务设施
Public Service Facilities

1990年代，公共服务设施不足百个，经多年发展，初步建成了较全面的服务体系。未来将加强精细化、差异化配置，建设国际化、高品质、全覆盖的高能级服务体系和多层次社区服务体系，形成安全、友好、活力、舒适的公共服务平台。

In the 1990s, there were less than 100 public service facilities in the region, while after years of development, a relatively comprehensive service system was initially established. In the future, it will strengthen refined and differentiated configuration, build an international, high-quality, universally available, high-level service system and a multi-level community service system, and form a safe, friendly, vigorous, and comfortable public service platform.

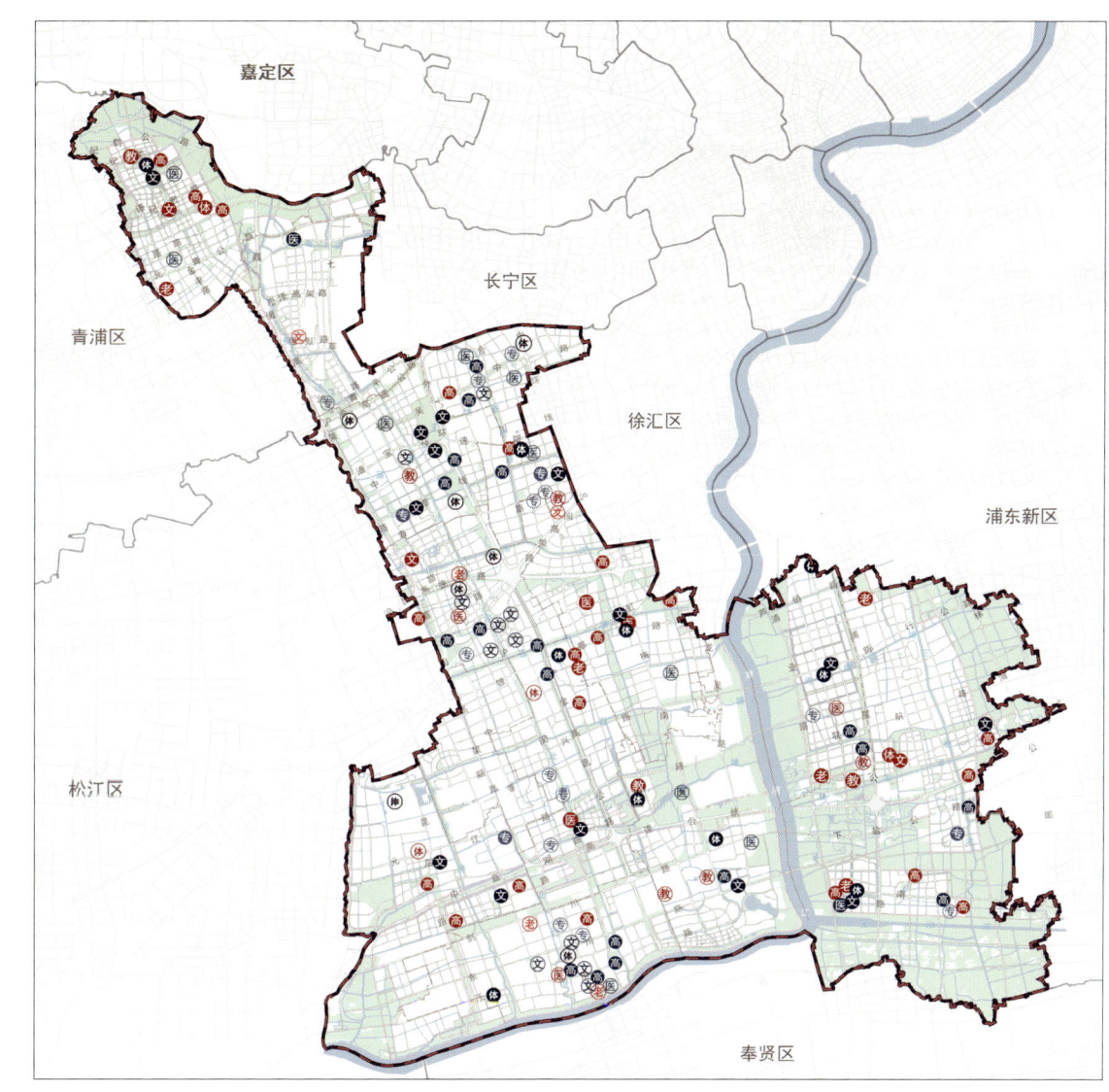

公共服务设施规划图（来源：闵行2035）
Public Service Facilities Planning (From: Minhang 2035)

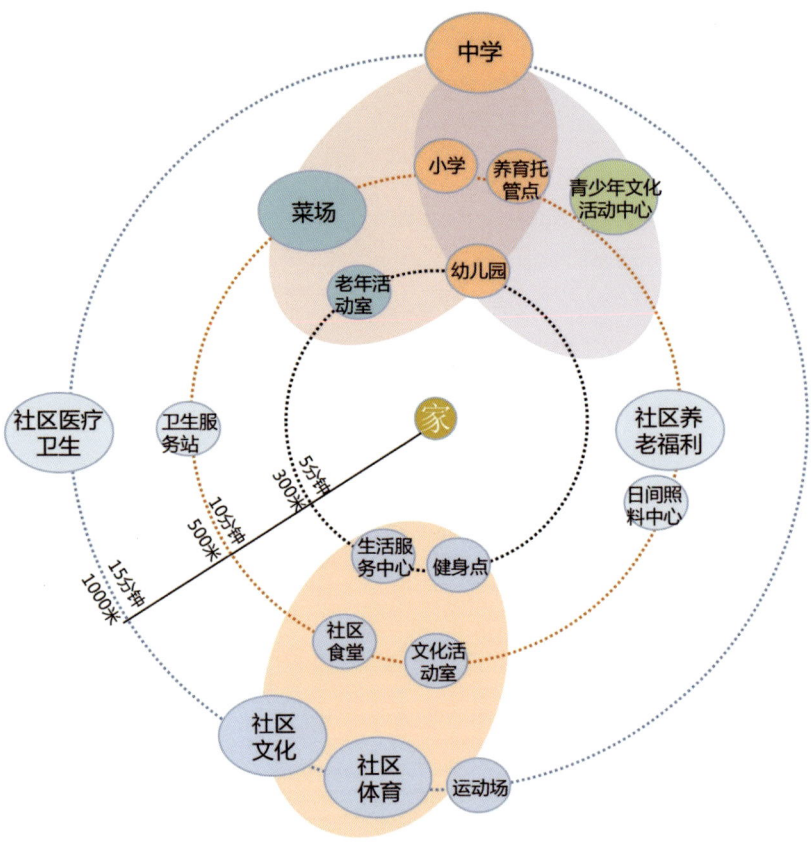

15分钟生活圈示意图
15-Minute Communite-life Circle

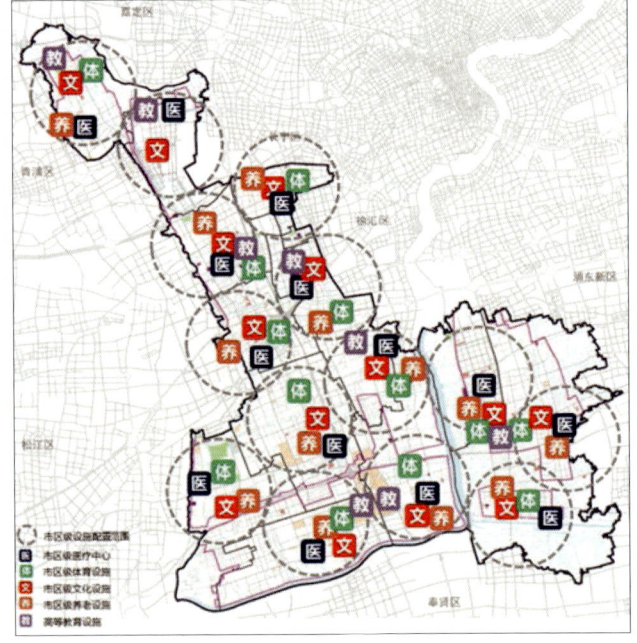

市区级设施每20万人全覆盖示意（来源：闵行2035）
Full Coverage of Urban Facilities for Every 200,000 People (From: Minhang 2035)

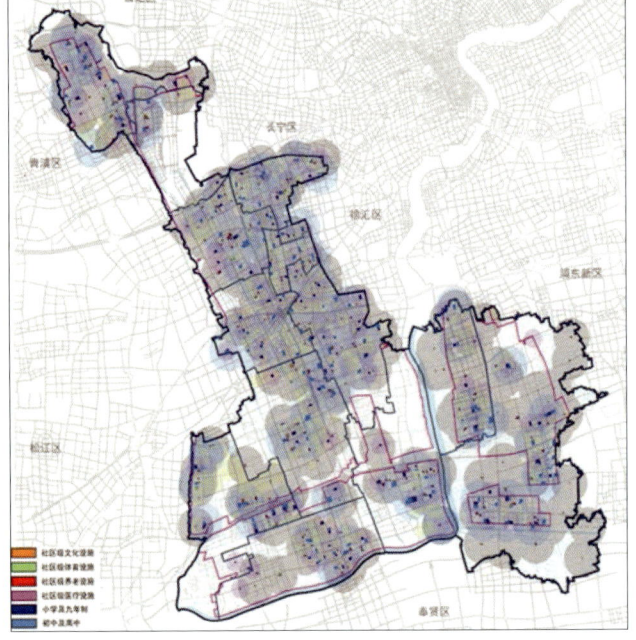

社区级设施15分钟生活圈全覆盖示意（来源：闵行2035）
Full Coverage of 15-Minute Communite-life Circle in Community-level Facilities (From: Minhang 2035)

新华书店
Xinhua Bookstore

社区菜场
Community Market

吴泾体育活动中心
Wujing Sports Center

第3章 社会民生 Part 3 Social Livelihood

钟书阁
Zhongshuge Bookstore

文化设施
Cultural Facilities

1990年代，仅有区级文化馆3个、图书馆2个、电影院5座。到2017年，共有市级文化设施1处、区级文化设施11处，以及多种类型的社区级文化设施。未来将新增3处市级、14处区级文化设施，以社区级公共服务设施15分钟步行覆盖率100%为目标，配置社区级文化设施。规划以高品质设施引领全区文化事业发展，丰富居民生活。

In the 1990s, there were only 3 district-level cultural centers, 2 libraries, and 5 movie theaters. By 2017, there was 1 municipal-level cultural facility, 11 district-level cultural facilities, and various types of community-level cultural facilities. In the future, 3 city-level and 14 districtlevel cultural facilities will be added, and community-level cultural facilities will be deployed so that any public service facility in communities can be reached in 15 minutes. It is planned to lead the development of cultural undertakings in the district with high-quality facilities and enrich residents' daily life.

宝龙美术馆
Powerlong Museum

义化设施规划图（来源：闵行2035）
Cultural Facilities Planning (From.Minhang 2035)

第3章 社会民生 Part 3 Social Livelihood

上海海派艺术馆
Shanghai Haipai Art Gallery

闵行博物馆新馆
New Minhang Museum

体育设施
Sports Facilities

1990年代，仅有闵行体育场、闵行游泳池、吴泾体育场3处成规模的体育场馆。随着城市的发展，相继建成上海射击活动中心、旗忠森林体育城网球中心等9处市区级体育设施，承办各项国际赛事，影响力不断扩大。未来将再建设10处市区级体育设施，满足各专业赛事的要求和居民对高品质体育设施的需求，并保障每个街镇至少有一处市民健身中心，使居民15分钟步行可达。

In the 1990s, there were only three large stadiums, Minhang Stadium, Minhang Natatorium, and Wujing Stadium. With the development of the city, 9 urban-level sports facilities such as the Shanghai Shooting & Archery Sports Center, Qizhong Forest Sports City Tennis Center, etc. have been successively built to host various international events, and their influence has been expanding. In the future, another 10 urban-level sports facilities will be constructed to meet the requirements of various professional events and residents' demand for high-quality sports facilities, and to make sure that each street or town has at least one fitness center within 15-minute walk from residents' community.

旗忠森林体育城网球中心（来源：马桥镇政府）
Qizhong Forest Sports City Tennis Center (From: Maqiao Town Government)

体育设施规划图（来源：闵行2035）
Sports Facilities Planning (From: Minhang 2035)

教育设施
Educational Facilities

1990 年代，仅有上海电机学院和上海交通大学两所高校，近年来随着华东师范大学、上海市电力工业学校、上海戏剧学院等高校相继入驻，高等教育水平得到了大幅提高。未来，根据城市副中心及新市镇建设要求，还将新建成 2—3 所专业性院校。

1990 年代，仅有 59 所基础教育学校。至 2018 年，已有 300 多所学校，并形成一批以七宝中学、莘松中学、明强小学等为代表的高水平教育设施。未来还将新增 21 所高中、31 所初中、11 所九年一贯制学校和 32 所小学，满足教育需求。

目前闵行已有上海美国学校、上海德威英国国际学校、星河湾双语学校等十余所国际学校，未来根据国际社区建设，将进一步完善国际学校布局，并依托南虹桥现有国际教育资源，打造国际教育特色功能区。

In the early 1990s, there were only two colleges, Shanghai Dianji University and Shanghai Jiao Tong University. In recent years, East China Normal University, Shanghai Electric Power Industry School, Shanghai Theater Academy, etc, settled in the district successively, which have largely improved the level of higher education. In the future, according to the construction requirements of sub-centers of city and new towns, 2–3 professional colleges will be built.

In the early 1990s, there were only 59 schools for basic education. By 2018, there were more than 300 schools, and a number of high-level educational facilities represented by Qibao High School, Xinsong Middle School, Mingqiang Elementary School, etc. In the future, 21 high schools, 31 junior high schools, 11 nine-year education schools and 32 primary schools will be added to meet the needs of education.

At present, there are more than 10 international schools, such as Shanghai American School, Dulwich College Shanghai (Minhang), Shanghai Star River Bilingual School, etc. In the future, the district will be developed into an international community by perfecting the layout of international schools, and meanwhile relying on the existing international education resources of Southern Hongqiao.

上海交通大学
Shanghai Jiao Tong University

上海交通大学
Shanghai Jiao Tong University

上海德威国际学校(来源:马桥镇政府)
Shanghai Dulwich College(From: Maqiao Town Government)

马桥文来外国语小学(来源:马桥镇政府)
Maqiao Wenlai Foreign Language Primary School (From: Maqiao Town Government)

医疗设施
Medical Facilities

1990年代初，仅有上海市第五人民医院、吴泾医院、昆阳医院等6所综合医院。经过多年发展，医疗设施规模及布局不断完善，相继引入复旦大学附属华山医院西院、复旦大学附属儿科医院、上海交通大学附属仁济医院等三甲医院。未来将改扩建3所、新增7所医院，并按照每5万—10万人配置1处社区卫生服务中心和若干卫生服务站，形成多层次、多类型的医疗卫生体系，让市民便捷就医、安全就医、有效就医。

In the early 1990s, there were only 6 general hospitals, including the Fifth People's Hospital of Shanghai, Wujing Hospital, Kunyang Hospital. After years of development, the scale and layout of medical facilities have been continuously improved, and the top three hospitals, such as the Huashan Hospital of Fudan University, Children's Hospital of Fudan University, and Renji Hospital of Shanghai Jiao Tong University, have been successively introduced. In the future, 3 hospitals will be rebuilt and expanded, 7 new hospitals will be built and 1 community health service center and a number of health service stations will be configured for every 50,000 to 100,000 people to form a multi-level and multi-type medical and health system, allowing citizens to seek medical treatment conveniently, safely and effectively.

闵行区中心医院
Minhang District Central Hospital

养老福利设施
Pension Welfare Facilities

伴随着日益增长的养老需求，相继建成了闵行区社会福利院、闵行区第二社会福利院、上海恩光敬老院等一批重要的养老设施。未来将建设老年友好型城市，规划新增5处市区级和12处社区级养老设施，形成社区为依托、机构为补充的养老服务体系，让老人安享多彩晚年。

Along with the increasing demand for old-age care, a number of important old-age care facilities have been built, such as Minhang Social Welfare Institute, Minhang No. 2 Social Welfare Institute, Shanghai Enguang Home for the Aged, etc. In the future, an elderly-friendly city will be built, and 5 urban-level and 12 community-level pension welfare facilities are planned to be added to form a community-based and institution-supplemented elderly care system to enable senior citizens to enjoy old age.

君莲养老院
Junlian Elderly Welfare Center

历史文化设施
Historical Sites & Cultural Facilities

闵行现存不可移动文物144处，其中已定级文物保护单位37处。未来将加强七宝老街、召稼楼老街等风貌保护街区及颛桥剪纸、马桥手狮舞等非物质文化遗产保护，留存本土文化印记，传承优秀传统文化。

There are currently 144 immovable cultural relics in Minhang, of which 37 have been classified as historical and cultural sites to be protected. In the future, the preservation of the styles of old streets, such as Qibao Old Street and Zhaojialou Old Street, and the protection of the intangible cultural heritage, such as Zhuanqiao Paper-cut and Maqiao Lion Dance, will be strengthened, in order to preserve the local cultural impact and inherit the excellent traditional culture.

闵行区市级以上文物保护单位
Municipal Cultural Relics Protection Unit

名称 Name	批次 Batch	公布日期 Date	所在街道 Street	年代 Age	类别 Category
马桥古文化遗址 Maqiao Ancient Cultural Site	第一批 First Batch	1977.12.7	马桥镇 Maqiao Town	新石器时代 Neolithic	古文化遗址 Ancient Cultural Site
上海县政府旧址 Former site of the Shanghai County Government	第八批 Eighth Batch	2014.4.4	颛桥镇 Zhuanqiao Town	1932	近现代重要史迹及代表性建筑 Important buildings of modern times
漕宝路七号桥碉堡 Caobao Road No. 7 Bridge Fort	第八批 Eighth Batch	2014.4.4	七宝镇 Qibao Town	民国 the Republic of China era	近现代重要史迹及代表性建筑 Important buildings of modern times
南张天主堂 Nanzhang Catholic Church	第八批 Eighth Batch	2014.4.4	莘庄镇 Xinzhuang Town	1901	近现代重要史迹及代表性建筑 Important buildings of modern times
上海普慈疗养院旧址 Site of Shanghai Mercy Hospital	第八批 Eighth Batch	2014.4.4	颛桥镇 Zhuanqiao Town	1935	近现代重要史迹及代表性建筑 Important buildings of modern times

上海普慈疗养院旧址
Site of Shanghai Mercy Hospital

漕宝路七号桥碉堡
Caobao Road No. 7 Bridge Fort

七宝天主堂
Qibao Catholic Church

南张天主堂
Nanzhang Catholic Church

第 4 章
生态休闲

PART 4
ECOLOGICAL LEISURE

生态网络体系
Eco-network System

闵行是我国首批国家级生态区，一直保持着良好的生态优势。目前已建成闵行体育公园、高压走廊绿化带、浦江生态林等生态空间，建立"点、线、面"结合的生态区。未来结合绿地、林地以及河湖水系等布局，衔接大治河生态走廊、近郊绿环、黄浦江间隔带，形成"一廊两环五带"的总体生态结构。

Minhang has always maintained a good ecological advantage and is the first batch of "National Eco-model Area" in China. At present, ecological spaces, such as Minhang Sports Park, Green Belts Corridor under high-tension cable, and Pujiang Ecological Forest, have been built, and an "ecological zone" combining "points (parks), lines (corridors), and areas (forests)" has been established. Combining the layout of green land, forest land, river and lake systems in the future, it will connect the Dazhi River Ecological Corridor, the suburban green belts, and the Huangpu River separation zone to form the overall ecological structure of "one corridor, two rings and five belts".

大治河生态走廊（来源：浦江镇政府）
Dazhi River Ecological Corridor (From: Pujiang Town Government)

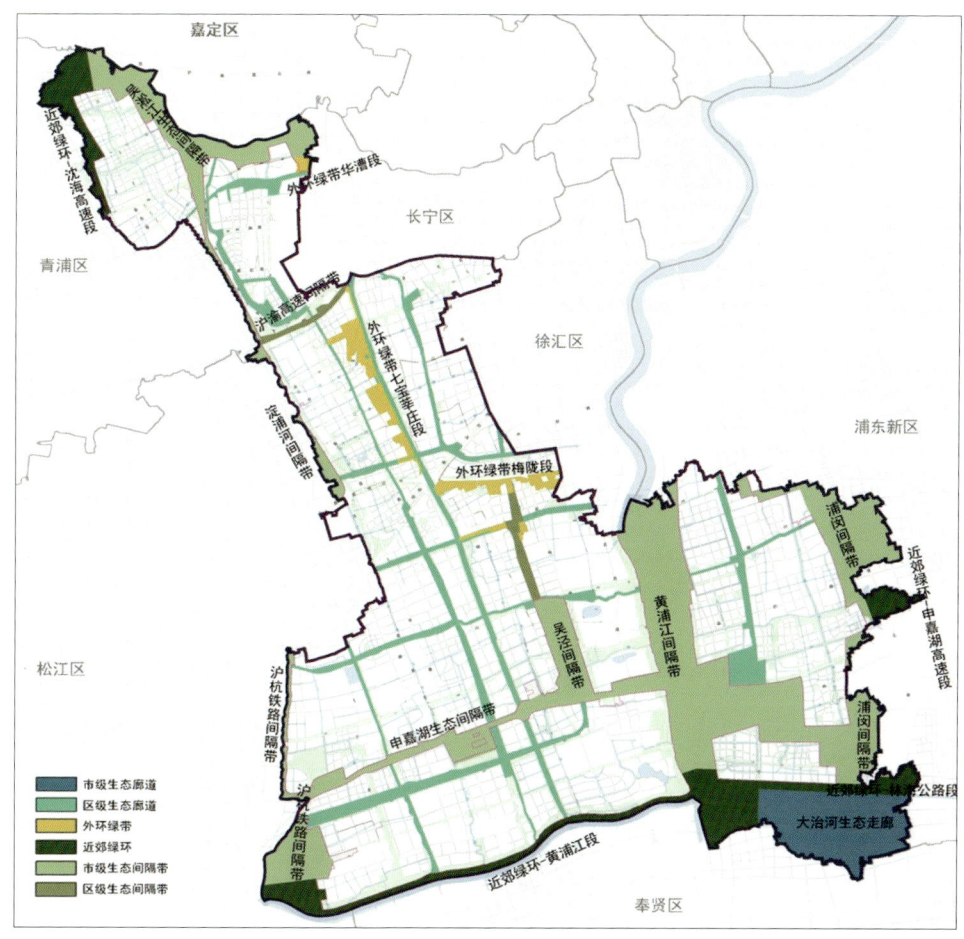

生态网络体系规划图（来源：闵行2035）
Ecological Network System Planning (From:Minghang 2035)

黄浦江生态间隔带
Huangpu River Separation Zone

第4章 生态休闲 Part 4 Ecological Leisure

河湖水系
Water System

河道纵横密布，拥有黄浦江、吴淞江（苏州河）、淀浦河等骨干河道 30 条段，300 多条支级河道。目前已形成浦西"七横四纵"、浦东"四横一纵"的骨干河道布局，河湖水面率为 8.2%，其中黄浦江在区内总长 18 公里。未来强化苏州河、大治河、春申塘等河道的保护与管控，修复骨干水体，恢复小型湖泊和河道水网，河湖水面率不低于 10.5‰。

The river channel is densely packed, with 30 main rivers including Huangpu River, Wusong River (Suzhou River), Dianpu River, etc. and more than 300 branch river channels. At present, the "seven crosswise and four lengthwise" river layout in Puxi and "four crosswise and one lengthwise" in Pudong have been formed. The water surface rate of rivers and lakes is 8.2%, of which the total length of the Huangpu River in the area is 18 kilometers. In the future, the protection and management of rivers, such as Suzhou River, Dazhi River and Chunshen Pond, will be strengthened, main water bodies will be repaired, and small lakes and river water networks will be restored, so that the water surface rate will not below 10.5%.

淀浦河黄昏
Dusk of Dianpu River

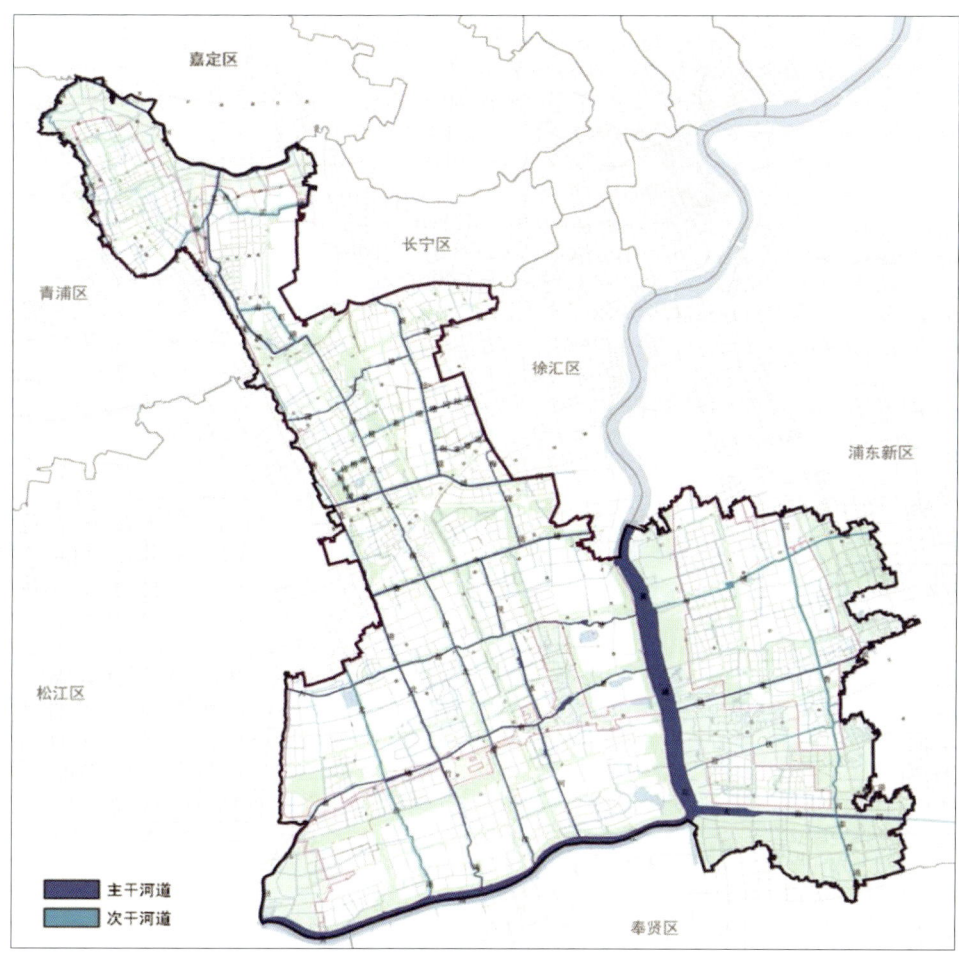

河湖水系规划图（来源：闵行2035）
Water System Planning (From:Minhang 2035)

北横泾
Beihengjing River

第4章　生态休闲　Part 4　Ecological Leisure

公园体系
Park System

1990年代初，公园个数少、规模小、分布散。随着城市的快速发展，目前已建成类型丰富、品质较高的各类公园，包括浦江郊野公园、闵行文化公园、莘庄城市公园等。未来将完善"郊野公园—城市公园—地区公园—社区公园—口袋（线性）公园"的五级城乡公园体系，人均公园绿地面积15.2平方米，实现出门见公园的目标。

In the early 1990s, there were only few small parks, scattered over the area. With the rapid development of the city, various types of parks with high quality have been built, including Pujiang Countryside Park, Minhang Culture Park, Xinzhuang Park, etc. In the future, a five-level urban-rural park system will be perfected, including country parks, city parks, regional parks, community parks, and pocket parks. In this way, the number of parks around communities can be largely increased, and the per capita park green area can reach 15.2 square meters.

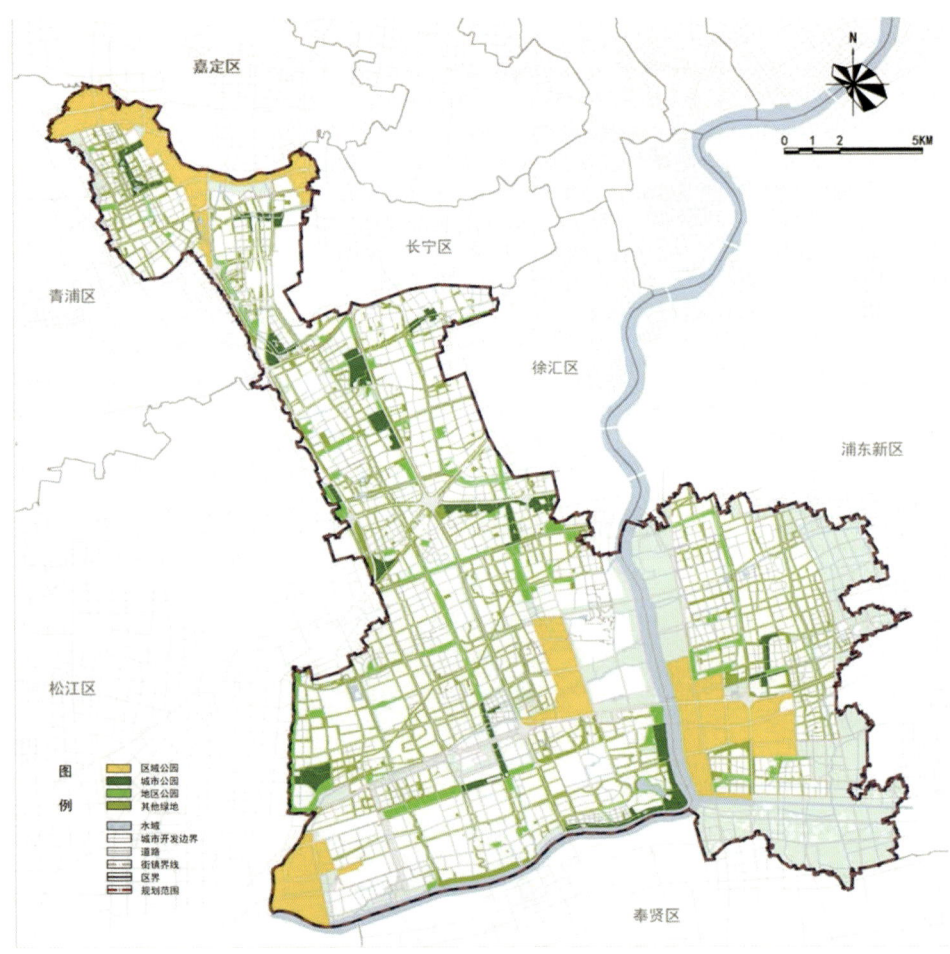

城乡公园体系规划图（来源：闵行2035）
Urban and Rural Park System Planning (From:Minhang 2035)

闵行文化公园
Minhang Cultural Park

浦江郊野公园
Pujiang Countryside Park

第4章 生态休闲 Part 4 Ecological Leisure

韩湘水博园
Hanxiang Water Expo Park

第4章 生态休闲 Part 4 Ecological Leisure

闵行体育公园
Minhang Sports Park

颛桥剪纸公园
Zhuanqiao Paper-cut Park

紫藤园
Wisteria Park

第4章 生态休闲 Part 4 Ecological Leisure

梅陇集心公园
Meilong Jixin Park

街头公园
Street Park

马桥古文化公园
Maqiao Ancient Culture Park

绿道体系
Greenway System

自2015年起，闵行大力推行绿道建设，卓有成效，已建成43条绿道，约107公里，包括滨江公园绿道、郊野公园绿道、沪闵路绿道等。未来将形成"区域级—城市级—社区级"三级绿道体系，绿道总长度不低于400公里，与公园体系协同发展，串联各个中心，贯通慢行系统，使城市建设与生态空间互动互融。

Since 2015, Minhang has vigorously promoted the construction of greenways, and has achieved great results. 43 greenways, about 107 kilometers, have been completed, including greenways in Binjiang Park, Countryside Park, and Humin Road. In the future, a three-level greenway system of "regional-urban-community-level" will be formed, and the total length of the greenway will be no less than 400 kilometers. It will develop in concert with the park system, connecting centers of each park as well as the slow traffic system to make urban construction interact and integrate with ecological space.

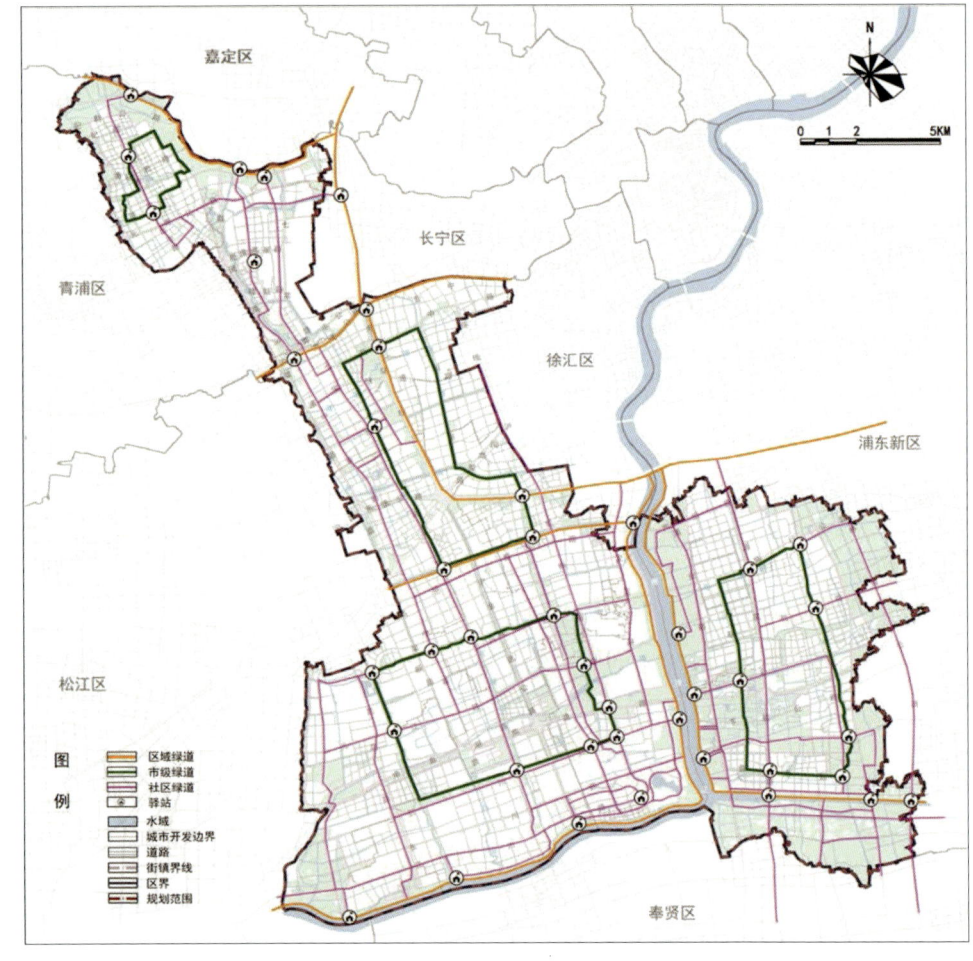

绿道系统规划图（来源：闵行2035）
Greenway System Planning (From: Minhang 2035)

闵行绿道
Minhang Greenway

闵行绿道
Minhang Greenway

第4章 生态休闲 Part 4 Ecological Leisure

休闲活动
Leisure Activities

古镇漫步
Sightseeing in Historical Towns

七宝老街、召稼楼等

闵行文化底蕴深厚，拥有许多远近闻名的古镇，既有江南水乡的自然风光，又有内涵深厚的历史文化积淀，每年都吸引着大量游客和当地居民前去观赏游览。

Qibao Old Street, Zhaojialou Ancient Town, etc.

Minhang has a long history and deep culture, with many well-known ancient towns. It has both the natural scenery of the Jiangnan and the profound historical and cultural background, which attracts a large number of tourists and local residents each year.

古镇游览（来源：闵行区规划资源局）
Historical Town Tour (From:Minhang Planning and Natural Resource Bureau)

艺术熏陶
Artistic Edification

宝龙美术馆、城市书房、海派艺术馆等

众多高品位文化艺术和娱乐空间，提供了重量级美术展览、近距离艺术体验、多剧种舞台演出等文化活动，让人们在文化和生活相融的热流中感受到高雅艺术的魅力。

Powerlong Museum, City Library, Haipai Art Gallery, etc.

Numerous high-grade cultural arts and entertainment spaces provide cultural activities such as art exhibitions, artistic experiences, multi-drama stage performances, etc., allowing people to feel the charm of elegant art in these places.

书房阅读（来源：闵行区规划资源局）
Reading in City Library (From:Minhang Planning and Natural Resource Bureau)

第4章 生态休闲 Part 4 Ecological Leisure

民俗体验
Folk-custom Arts

颛桥剪纸、七宝皮影戏等

马桥手狮舞、华漕小锣鼓等地方民俗已在闵行流传数百年，其中有些已被列入非物质文化遗产。如今，各种展示及体验活动让人们近距离全面、生动地了解传统民俗文化。

Zhuanqiao Paper-cut, Qibao Shadow Play, etc.

Local folk customs such as Maqiao Lion Dance, Huacao Small Gong and Drum, etc. have been inheriting in Minhang for hundreds of years, and some of them have been listed as intangible cultural heritage. Today, various exhibitions and experience activities allow people to get a comprehensive and vivid understanding of traditional folk culture closely.

手狮舞表演（来源：马桥镇政府）
Lion Dance Show (From: Maqiao Town Government)

美食盛宴
Gourmet Banquets

士林夜市、慕尼黑啤酒节、十尚坊美食节等

各种美食节和美食街颇受游客和当地居民的欢迎，品特色小吃、喝香醇美酒、赏精彩演出，热闹欢快的现场气氛体现了丰富多彩、动感十足的城市活力。

Shilin Night Market, Munich Oktoberfest, The Ten Gourmet Festival, etc.

Various food festivals and food streets are popular with tourists and local residents, who can enjoy specialty snacks, drink delicious wines, and enjoy wonderful performances. The lively atmosphere reflects the colorful and dynamic city vitality.

美食狂欢
Gourmet Carnival

运动盛会
Sporting Events

定向骑行、马拉松等

闵行每年都会举办多场公益骑行及马拉松赛事，营造了热爱运动、坚持挑战的生活氛围，提高了人们关注运动、参与运动、全民健身的意识。

Cycling, Marathon, etc.

Minhang hosts several public-spirited cycling and marathon events each year, creating an atmosphere where people love sports and persist in challenges, thereby raising people's awareness of national fitness by understanding sports and participating in sports.

万科城市乐跑
Vanke City Run for Fun

公益参与
Public Benefit Activities

志愿服务等

许多公益团队组织各项公益活动,如学雷锋志愿服务、进社区关爱老年人行动、义务植树活动等,推动了社区公益项目多元化专业化发展和社会公益氛围的形成。

Volunteer Services

Many public volunteer teams organize various activities, such as providing voluntary service, caring for the elderly in communities, tree-planting activities, etc., which promote the diversified and professional development of community public welfare programs and the formation of an atmosphere in which people are willing to offer a helping hand to others.

义务植树
Voluntary Tree-planting

第 5 章
乡村蜕变

PART 5
RURAL METAMORPHOSIS

乡村发展
Rural Development

随着城市化进程的快速推进，村庄数量逐年递减，行政村由 1992 年的 170 个减少到 2017 年的 122 个，目前村庄主要位于华漕镇、吴泾镇、马桥镇、浦江镇和浦锦街道。为了配合城市建设和生态网络空间等重大战略落实，未来将继续优化农业布局，引导农村居民点集中归并，规划到 2035 年保留保护行政村 10 个，农村人口约 1.3 万人，保留农村居民点用地约 1.6 平方公里。

With the rapid progress of urbanization, the number of villages has been decreasing year by year. The number of administrative villages has decreased from 170 in 1992 to 122 in 2017. At present, the villages are mainly located in Huacao Town, Wujing Town, Maqiao Town, Pujiang Town and Pujin Street. In order to cooperate with the implementation of major strategies such as urban construction and ecological cyberspace, the agricultural layout will continue to be optimized in the future, leading to the centralized consolidation of rural settlements. It is planned to retain and protect 10 administrative villages by 2035, with a rural population of about 13,000, and retain 1.6 square kilometers agricultural land.

闵行区现状村庄分布图（来源：上海市闵行区村庄布局规划（2017—2035））
The Distribution Map of Present Villages in Minhang District
(From: Planning of Village Distribution in Minhang District, Shanghai (2017-2035))

闵行区规划村庄分布图（来源：上海市闵行区村庄布局规划（2017—2035））
The Distribution Map of Planning Villages in Minhang District
(From: Planning of Village Distribution in Minhang District, Shanghai (2017-2035))

浦江镇革新村
Gexin Village of Pujiang Town

农业生产
Agricultural Production

从1990年代中后期开始，闵行确立"都市农业"的发展目标。到2018年，全年粮食播种面积1.83万亩（约12.2平方公里），蔬菜生产面积0.96万亩（约6.4平方公里），农业规模化生产比例达90%以上。未来闵行将积极打造与美丽乡村融为一体的农业环境，提供生态型优质农产品和衍生服务产品，展现精品农业新形象。

Since the mid-to-late 1990s, Minhang has established the development goal of "urban agriculture". By 2018, the total area sown to grain will be 18,300 mu (12.2 km^2), and the land for vegetable production will be 9,600 mu (6.4 km^2). The proportion of agriculture mass production will reach 90%. In the future, Minhang will actively create an agricultural environment integrated with beautiful villages, provide ecological high-quality agricultural products and derivative products, and display a new image of high-quality agriculture.

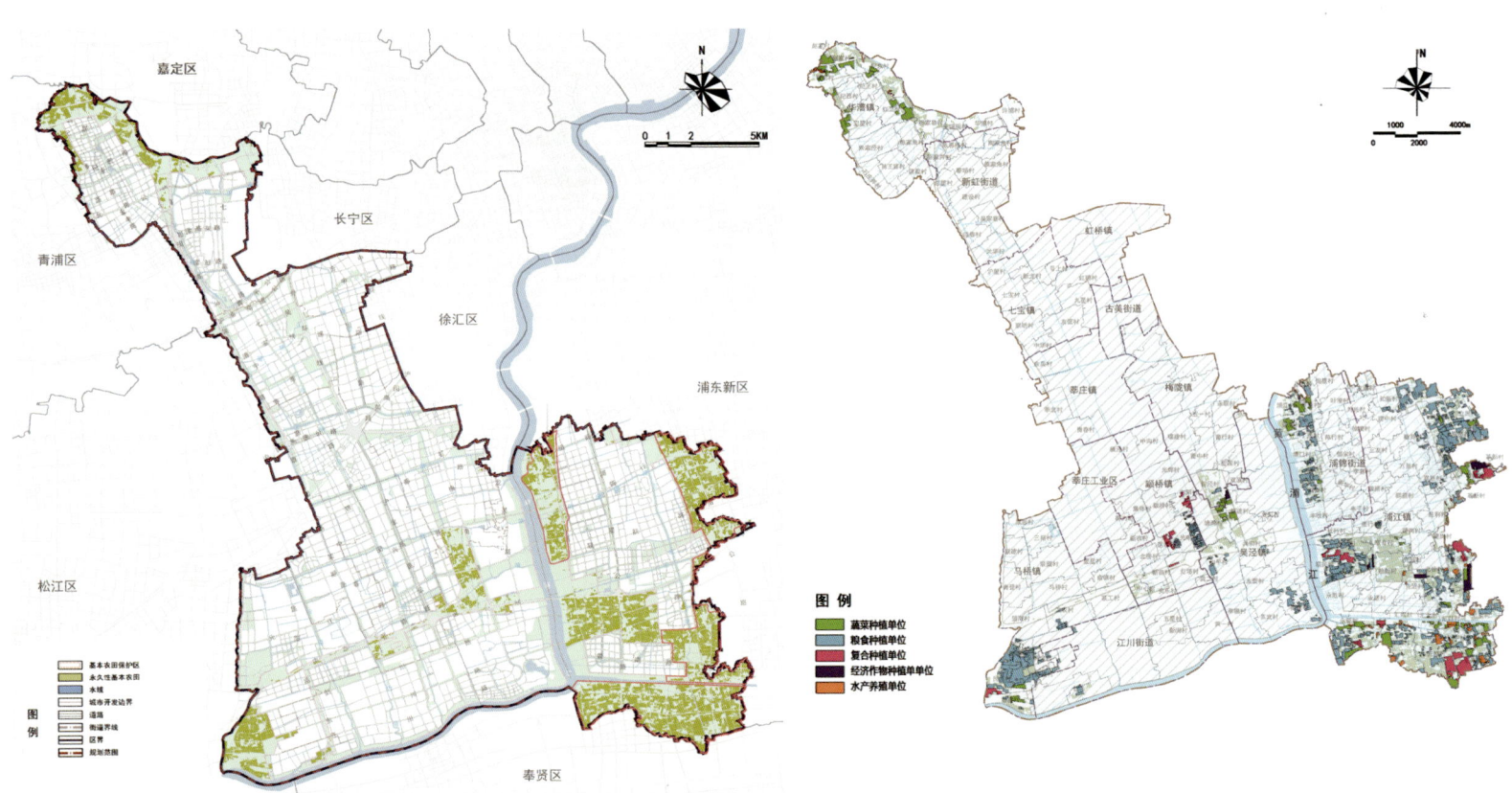

闵行区永久基本农田规划图（来源：闵行2035）
Permanent Basic Farmland Planning (From:Minhang 2035)

农业设施用地布局现状图（来源：闵行区农业布局规划）
Current Situation of Agricultural Facility Land Layout (From: Agricultural Layout Planning of Minhang District)

马桥农田收获前夕（来源：马桥镇政府）
Harvest in the Farmland of Maqiao (From: Maqiao Town Government)

马桥农田收获前夕（来源：马桥镇政府）
Harvest in the Farmland of Maqiao (From: Maqiao Town Government)

乡村生活
Country Life

近年来，闵行新农村建设相继开展，农民生活水平有显著改善，重点布局"三室两点"（村委会办公室、文化活动室、医疗诊治室、村民健身点和便民点）。未来将以改善农村居住环境，推动乡村振兴为基本原则，建设美丽乡村，完善农村公共服务设施和基础设施，增强村民获得感和幸福感。

In recent years, the construction of new rural areas in Minhang has been successively carried out, and the living standards of farmers have significantly improved. The focus is on "Three Rooms & Two Centers" (village committee office, cultural activity room, medical room, village fitness center and convenience services center). In the future, the basic principle is to improve the rural living environment and promote the rural vitalization, build beautiful village, improve rural public service facilities and infrastructure, and enhance villagers' sense of well-being and happiness.

"三室两点"建设
Construction of "Three Rooms & Two Centers"

村落农宅
Village Farmhouse

乡村地区多依托河道、湖泊等水系形成临河布局的带状村落。因村组规模小、建房管理粗放等历史原因，现状村宅布局较为分散。未来将对保留村庄内村宅进行逐步归并，撤并零散居民点和位于生态敏感区的村庄，按照节约集约用地、保护乡土文化特色等原则，系统性优化乡村居住社区，推进农民相对集中居住。

Rural areas rely mostly on rivers, lakes and other water systems to form strip-shaped villages that are adjacent to the river. Due to historical reasons such as the small size of the village groups and extensive construction management, the current layout of village houses is relatively scattered. In the future, the village houses in the reserved villages will be gradually merged, and scattered settlements and villages located in ecologically sensitive areas will be merged. In accordance with the principles of economical and intensive land use and protection of local cultural characteristics, the rural residential communities will be systematically optimized so that farmers will live less dispersedly.

带状村落
Strip-shaped Village

农居点意向
Intention of Agricultural Settlement

田园风貌
Landscape in Countryside

"芳草鲜美，落英缤纷，土地平旷，屋舍俨然"，在闵行，有许多和自然风光融为一体的田园乡村，主要分布在浦江镇和马桥镇。伴随着乡村振兴战略的实施和美丽乡村的全面建设，未来闵行的乡村将粉墙黛瓦、水清岸绿、枕河而居。

"The fragrant herbage was fresh and beautiful; fallen blossom lay in profusion. And on the flat area there were houses of a stately appearance." In Minhang, there are many rural villages integrated with natural scenery, mainly distributed in towns like Pujiang Town and Maqiao Town. With the implementation of the strategy of rural vitalization and the construction of beautiful villages in all aspects, the villages of Minhang will have neat houses, clear water, and green environment in the future.

田园风光（来源：马桥镇政府）
Rural Scenery (From: Maqiao Town Government)

美丽乡村
Beautiful Villages

为贯彻中央新型城镇化的发展要求，自 2016 年起闵行实施"美丽乡村"建设，农村居住环境得到明显改善。2018 年开始，以"美丽家园、绿色田园、幸福乐园"为抓手，实施乡村振兴战略，构建高水平的城乡发展一体化新格局。随着美丽乡村建设的进一步完善，未来将打造更多乡村振兴示范村，留住美好乡愁。

In order to implement the requirements of New-type Urbanization put forward by the central government, Minhang has implemented the construction of "Beautiful Village" since 2016, and the rural living environment has been significantly improved. From 2018, with "beautiful homes, green farmlands, and happy life" as the starting point, Minhang has been implementing the rural revitalization strategy and building a new pattern with a high-level urban-rural development integration. With the further improvement of the construction of beautiful villages, more rural demonstration villages will be built in the future to make rural areas more attractive.

美丽乡村（来源：马桥镇政府）
Beautiful Village (From: Maqiao Town Government)

第5章 乡村蜕变 Part 5 Rural Metamorphosis

马桥镇彭渡村
Pengdu Village of Maqiao Town

彭渡村（又称荷巷桥），位于马桥镇西南角，被列入全国第一批中国传统村落名录，村域面积约 1.8 平方公里。村内有一条荷巷桥老街，兴起于清朝嘉庆年间，村名逐渐改称为荷巷桥，村落内有 4 处不可移动文物和多处历史建筑。未来彭渡村将以传统村落保护为主体，以文化体验为目的，打造都市近郊集休闲农业、居住、集市于一体的传统风貌保护区。

Pengdu Village (also known as Hexiangqiao), located in the southwest corner of Maqiao Town, is listed in the first batch of China traditional villages list. The area of the village is about 1.8 square kilometers. There is an old street named Hexiangqiao in the village, which emerged during the Jiaqing years of the Qing Dynasty, and gradually the village changed its name to Hexiangqiao. There are 4 immovable cultural relics and many historical buildings in the village. In the future, Pengdu Village will take the protection of traditional villages as the main body and the cultural experience as the purpose to create a conservation area with traditional features in the suburbs of the city by integrating leisure agriculture, residence and markets.

顾言故居
Gu Yan's Former Residence

金氏宗祠
Jin's Ancestral Hall

荷巷桥老街
Hexiangqiao Old Street

金庆章故居
Jin Qingzhang's Former Residence

金家住宅仪门
Secondary Gate of Jin's Residence

整体效果图（来源：闵行区马桥镇彭渡村传统村落保护与发展规划）
Overall Rendering (From: Protection and Development Planning of Pengdu Traditional Village in Maqiao Town, Minhang District)

第5章　乡村蜕变　Part 5　Rural Metamorphosis

浦江镇革新村
Gexin Village of Pujiang Town

革新村位于浦江镇东部，是召稼楼古镇所在地，村域面积约 2.4 平方公里。作为中国传统村落和中国历史文化名村，有街巷、河道、桥梁、古树、水井、河埠、特色驳岸及特色铺地等历史环境要素，全村范围内有区级文物保护单位 3 处、文物保护点 6 处。"十里晓烟破，数声召稼钟"，依托召稼楼古镇旅游资源，集古镇旅游、农耕休闲、踏青度假于一身的革新村，确定了"田园文化旅游示范村"的发展定位，于 2019 年成为上海市首批、闵行区首个乡村振兴示范村。

Gexin Village is located in the east of Pujiang Town and is the site of Zhaojialou Ancient Town. The village is about 2.4 square kilometers. As a traditional Chinese village and a famous historical and cultural village in China, there are historical environmental elements such as streets, rivers, bridges, ancient trees, water wells, river ports, unique revetments and ancient roads. There are 3 historical and cultural sites protected at the district level within the village, and 6 cultural relics protection points. Based on the tourism resources of Zhaojialou Ancient Town, Gexin Village has established the development orientation of the "Demonstration Village of Pastoral Cultural Tourism". In 2019, it is listed in the first batch of demonstration villages in rural revitalization in Shanghai, becoming the first village awarded in Minhang District.

召稼楼古镇
Zhaojialou Ancient Town

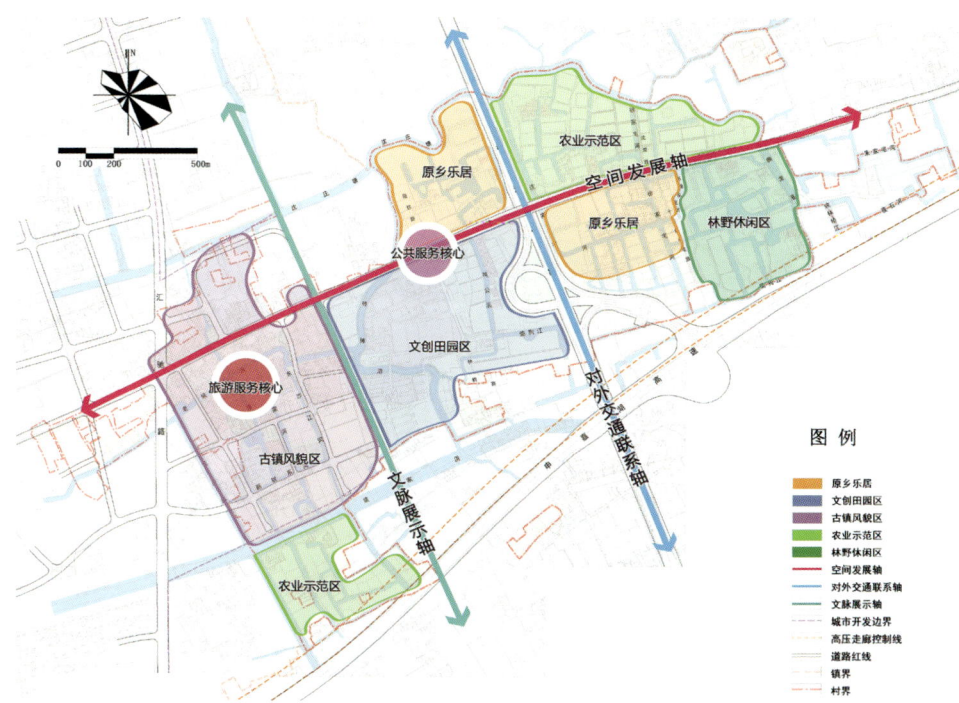

空间结构规划图（来源：闵行区浦江镇革新村传统村落保护与发展规划）
Space Structure Planning (From: Protection and Development Planning of Gexin Traditional Villages in Pujiang Town, Minhang District)

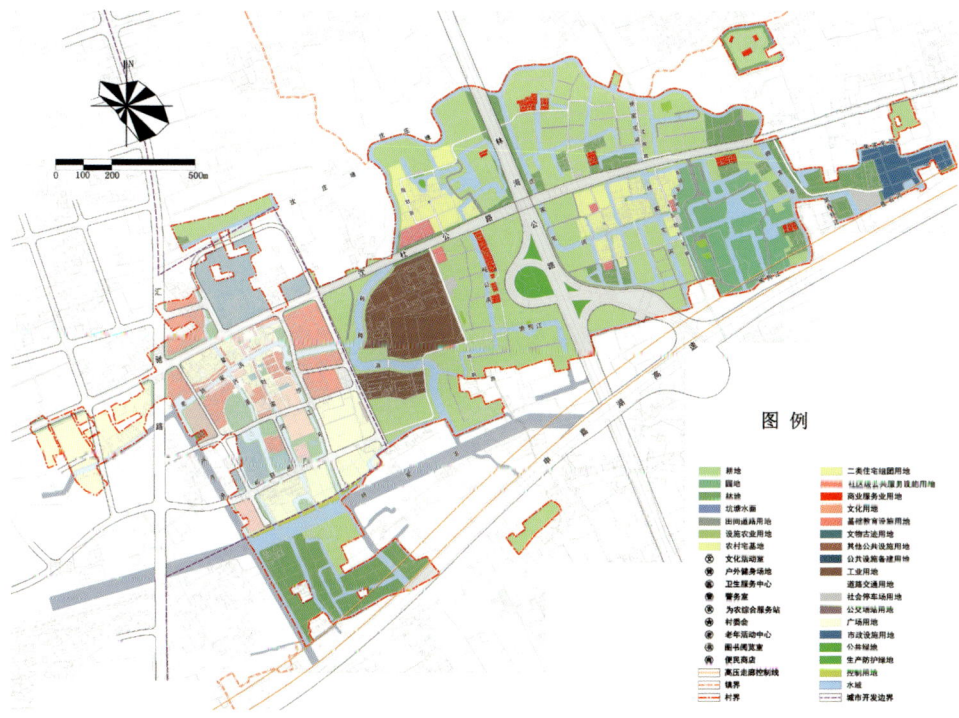

用地规划图（来源：闵行区浦江镇革新村传统村落保护与发展规划）
Land Use Planning (From: Protection and Development Planning of Gexin Traditional Villages in Pujiang Town, Minhang District)

第5章 乡村蜕变 Part 5 Rural Metamorphosis

召稼楼古镇
Zhaojialou Ancient Town

梅陇镇许泾村
Xujing Village of Meilong Town

许泾村地处梅陇镇最南端,全村总面积 2.2 平方公里。2016 年许泾村"美丽乡村"建设正式启动,通过"5+1"工程,即硬件(绿、路、水、电、房)和软件(文化客堂间)全方位改造和建设,村庄环境得到很大改善,许泾村走上了一条可持续的生态发展之路。

Xujing Village is located at the southern end of Meilong Town, with a total area of 2.2 square kilometers. In 2016, the construction of the "Beautiful Village" in Xujing Village was officially launched. Through the comprehensively transformation and construction of the "5+1" project, that is, the hardware (green, road, water, electricity, housing) and software (Culture Ketangjian), the environment of the village has been largely improved and the village has embarked on a sustainable ecological development road.

许泾村
Xujing Village

许泾村
Xujing Village

第5章 乡村蜕变 Part 5 Rural Metamorphosis

第 6 章
地区风采

PART 6
REGIONAL
ELEGANCE

虹桥商务区华漕镇
Hongqiao Central Business District of Minhang (Huacao Town)

区位
Location

虹桥商务区的城市功能核心，辖区面积约28.2平方公里，常住人口23.4万人。由原闵北工业区转型升级为现代商务区，通过区域整体优化，加快推进成片动迁腾地、同步实施基础建设、建设新虹桥国际医学中心、打造优质教育资源。未来以前湾公园为核心，配置国际服务、国际医疗、国际教育、国际创新、国际文化、国际居住六大功能，打造卓越全球城市的样板区。

The core of urban function of Hongqiao Business District is about 28.2 square kilometers, with a resident population of 234,000. The former Minbei Industrial Zone was transformed and upgraded to a modern business unit. Through the overall optimization of the Southern Hongqiao area, it has accelerated the relocation of land, simultaneously implemented infrastructure, built a new Hongqiao International Medical Science Center, and created high-quality educational resources. In the future, Qianwan Park will be the core and will be equipped with six functions, international services, international medical treatment, international education, international innovation, international culture, and international residence, to create a model of outstanding global cities.

南虹桥地区城市设计效果图
Urban Design Rendering of Southern Hongqiao Area

用地规划图（来源：华漕镇已批控制性详细规划拼合）
Land Use Planning (From: Control Detail Plan of Huacao Town)

第6章 地区风采　Part 6 Regional Elegance

国际医学中心园区
International Medical Center Park

华漕国际社区文化活动中心
Huacao Community Center

美丽乡村赵家村
Beautiful Zhaojia Village

虹桥商务区新虹街道
Hongqiao Central Business District of Minhang (Xinhong Street)

区位
Location

新虹街道是虹桥商务区的商务功能核心区域，辖区面积约19.3平方公里，常住人口约6万人。未来将充分发挥交通枢纽和商务功能的集聚整合作用，建设功能多元、交通便捷、空间宜人、生态高效、极强发展活力和吸引力的上海首个低碳商务社区。

The community is located in the core area of the Hongqiao Business District, with an area of about 19.3 square kilometers and a resident population of about 60,000. In the future, it will give full play to the integration of transportation hubs and business functions, and build Shanghai's first low-carbon business community with multiple functions, convenient transportation, pleasant space, ecological efficiency, and strong development vitality and attractiveness.

新虹街道（来源：新虹街道办事处）
Xinhong Street (From: Xinhong Street Office)

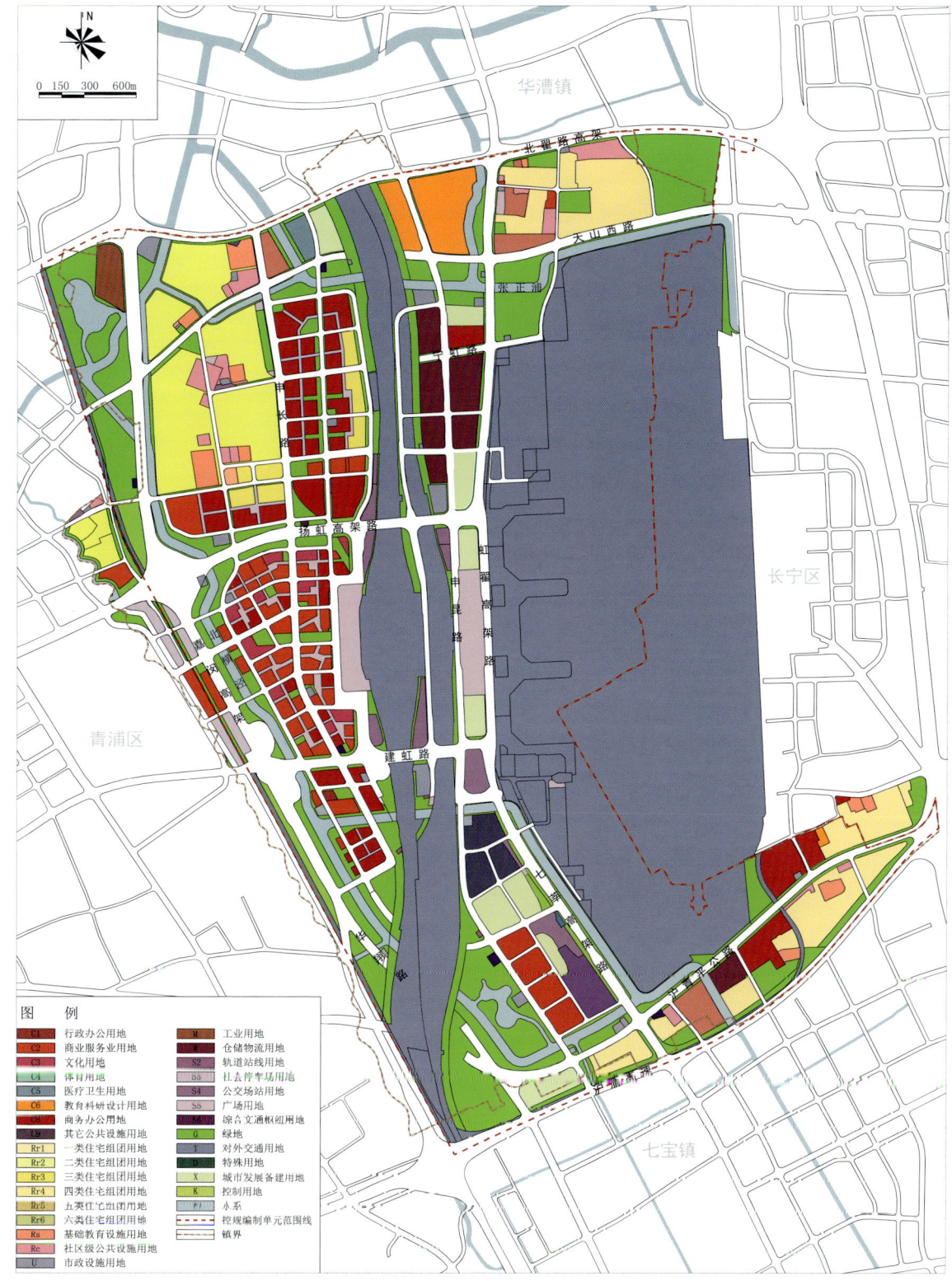

用地规划图（来源：新虹街道已批控制性详细规划拼合）
Land Use Planning (From: Control Detail Plan of Xinhong Street)

商务区
Business Area

高架与天桥
Elevated Road & Overpass

北横泾
Beihengjing River

第6章 地区风采 Part 6 Regional Elegance

七宝镇
Qibao Town

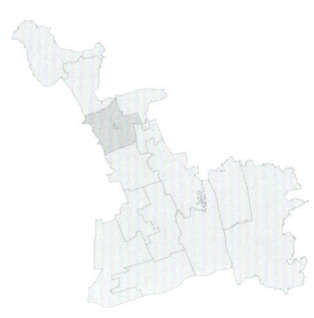

区位
Location

古镇七宝，辖区面积约 20 平方公里，常住人口 30.5 万人，镇内七宝老街是上海历史文化风貌区之一。近年来，依托七宝生态商务区等重点项目，加快经济转型提升，同时建成以闵行文化公园为代表的蓝绿生态网络，人居环境更加优美。未来将以九星地区综合开发和机场联络线七宝站建设为重要引擎，积极打造联系大虹桥、大浦东的桥头堡，助力闵行地区经济高质量发展。

Qibao Ancient Town, with an area of 20 square kilometers and a resident population of 305,000, is one of the historical and cultural districts in Shanghai. In recent years, relying on key projects such as the Qibao Ecological Business District it has accelerated economic transformation and improvement. At the same time, it has built a blue-green ecological network represented by Minhang Cultural Park, which has a more beautiful living environment. In the future, the comprehensive development of Jiuxing area and the construction of Qibao Station on the terminal connecting line will be used as important engines to actively build bridgeheads connecting Great Hongqiao and Great Pudong to help the high-quality economic development in Minhang.

七宝古镇
Qibao Ancient Town

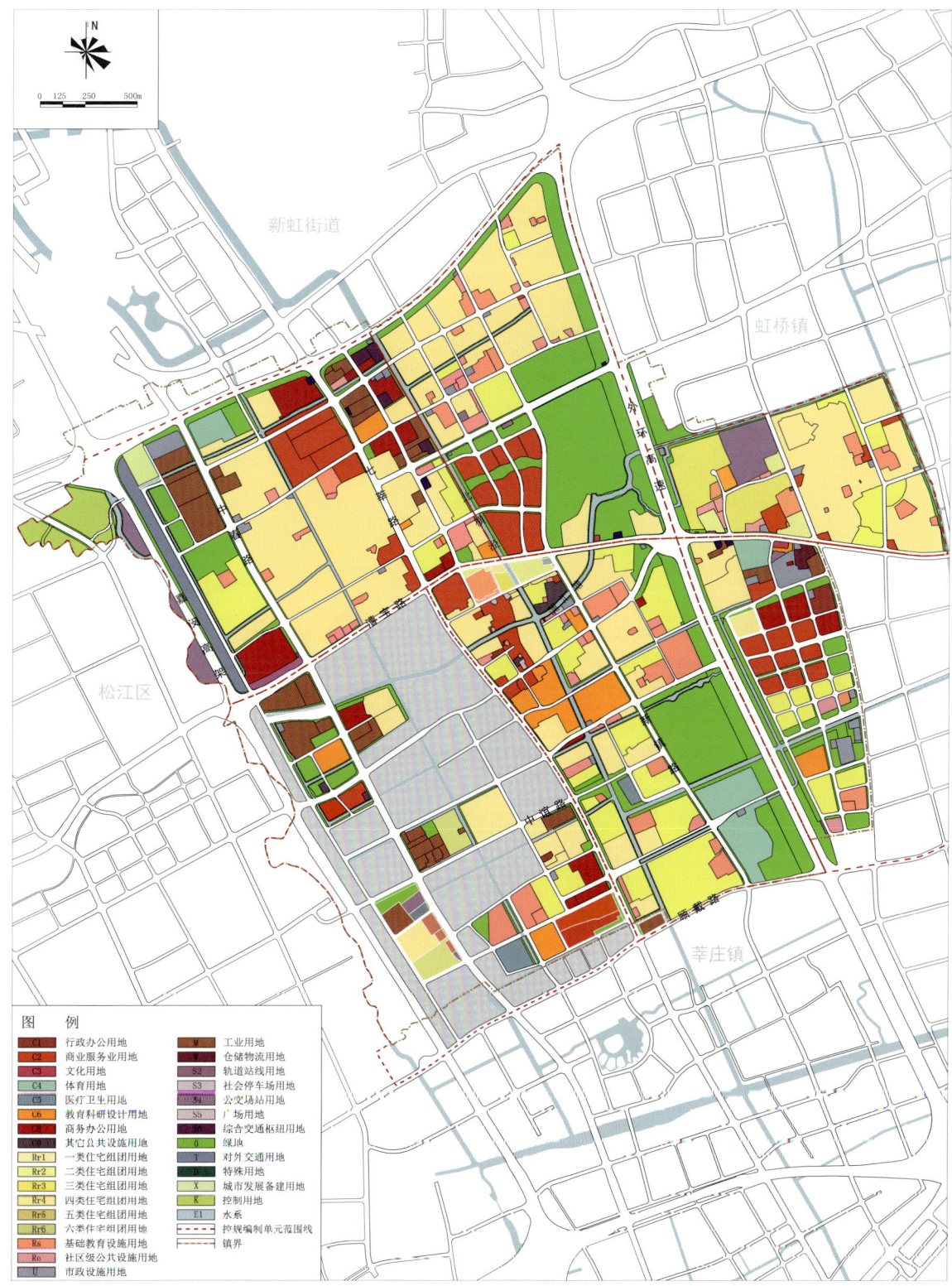

用地规划图（来源：七宝镇已批控制性详细规划拼合）
Land Use Planning (From: Control Detail Plan of Qibao Town)

第6章 地区风采 Part 6 Regional Elegance

七宝商务区
Qibao Business District

七宝老街
Qibao Old Street

闵行文化公园
Minhang Cultural Park

第6章　地区风采　Part 6 Regional Elegance

虹桥镇
Hongqiao Town

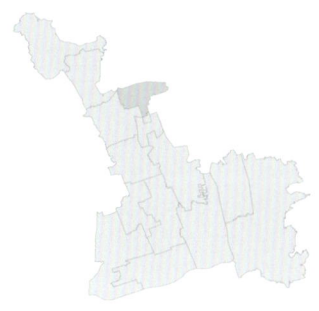

区位
Location

虹桥镇因"蒲汇塘上有石桥，雨过天晴见彩虹"而得名。位于闵行区中东部，毗邻长宁、徐汇，是中心城内的镇，区位优越，辖区面积约11平方公里，常住人口约15.3万人。20世纪90年代，随着房地产的发展和人口大量导入，地区经济实现快速发展。紧临古北，外籍人口多，国际社区氛围浓郁，现有爱琴海、万象城、高尔夫球场等大型商业综合体和文体设施。未来的虹桥镇将建成品质卓越、生态宜居的现代化新城区。

Hongqiao Town is an international comprehensive community with an area of about 11 square kilometers and a resident population of about 153,000. Hongqiao Town, literally rainbow bridge, gets its name because there is a stone bridge over Puhuitang, the Puhuitang Bridge, and when the sun shines again after rain, people can see rainbows there. In the 1990s, with the development of real estate and the large-scale introduction of population, the regional economy achieved rapid development. There is a large foreign population, creating a strong atmosphere of international community. There are large commercial complexes and cultural and sports facilities such as the Aegean Shopping Park, the Mixc Shopping Center, and golf courses. In the future, Hongqiao Town will be built into a modern new town of excellent quality and ecological livability.

虹桥镇（来源：虹桥镇政府）
Hongqiao Town (From: Hongqiao Town Government)

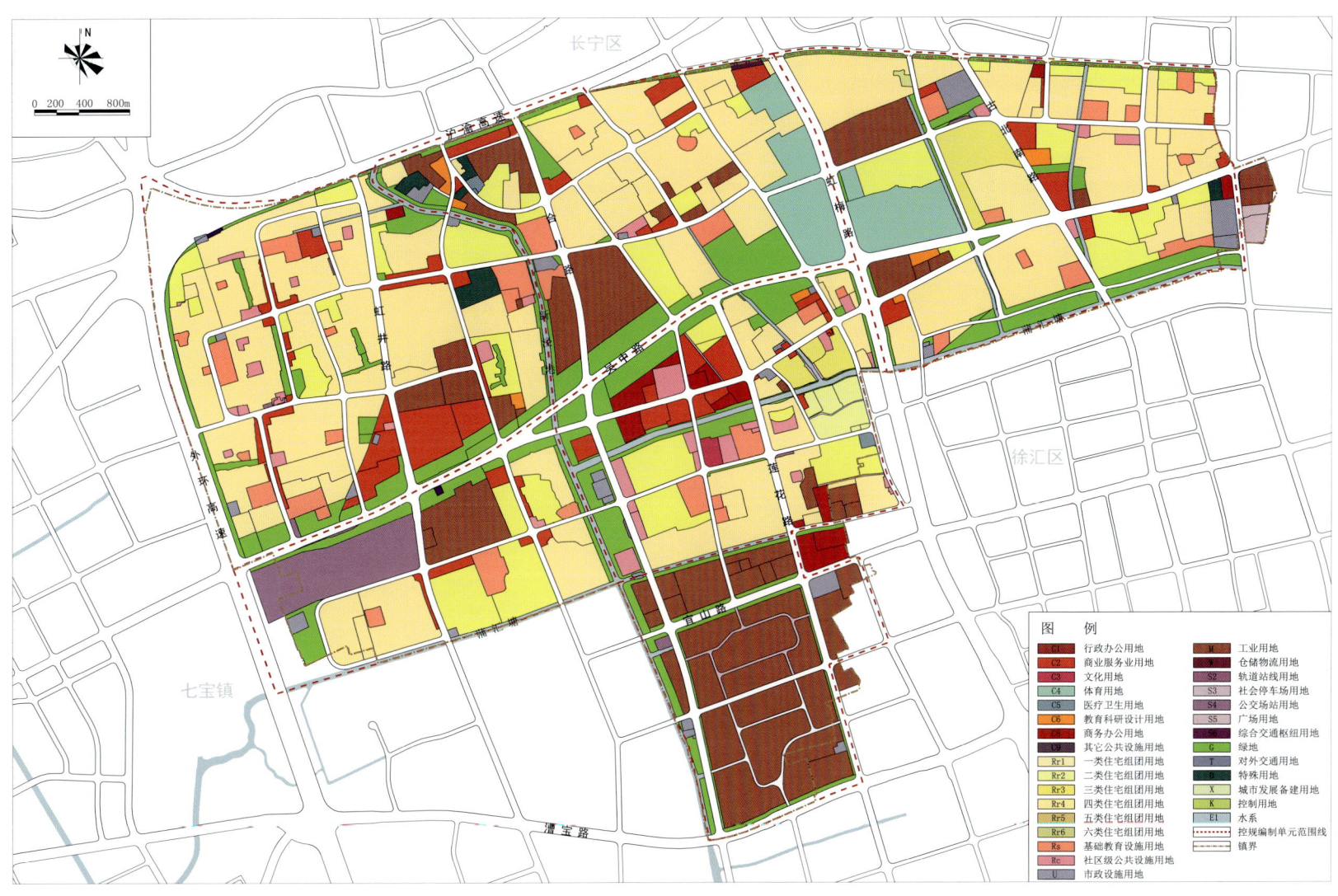

用地规划图（来源：虹桥镇已批控制性详细规划拼合）
Land Use Planning (From: Control Detail Plan of Hongqiao Town)

第6章 地区风采 Part 6 Regional Elegance

万象城（来源：虹桥镇政府）
The Mixc Shopping Center (From: Hongqiao Town Government)

老外街夜景（来源：虹桥镇政府）
Night View of Foreign Street (From:Hongqiao Town Government)

阿拉城（来源：虹桥镇政府）
Ala Town (From:Hongqiao Town Government)

古美路街道
Gumei Street

区位
Location

　　古美路街道是宜居型的高品质综合社区，辖区面积6.5平方公里，常住人口11.2万人。自1999年从梅陇镇析出以来，依托1号线莲花路、外环路站点的区位优势，迅速成长为一个交通便捷、购物便利、公共设施日趋完善、发展潜力相当大的新建住宅区与人口导入区。未来，古美路街道将继续秉承"和谐古美、品质生活"的发展愿景，向集商贸、购物、文化、娱乐于一体的综合性社区转变。

Gumei Street is a high-quality, livable community with an area of 6.5 square kilometers and a resident population of 112,000. Since its separation from Meilong Town in 1999, it has rapidly grown into a new residential area with convenient transportation, convenient shopping, improved public facilities, and considerable development potential, relying on the location advantages of Metro Line 1 Lianhua Road and Outer Ring Road Station. In the future, the Street will continue to uphold the development vision of "harmonious Gumei with high quality life" and transform into a comprehensive community integrating commerce, shopping, culture, and entertainment.

古美路街道（来源：古美路街道办事处）
Gumei Street (From: Gumei Street Office)

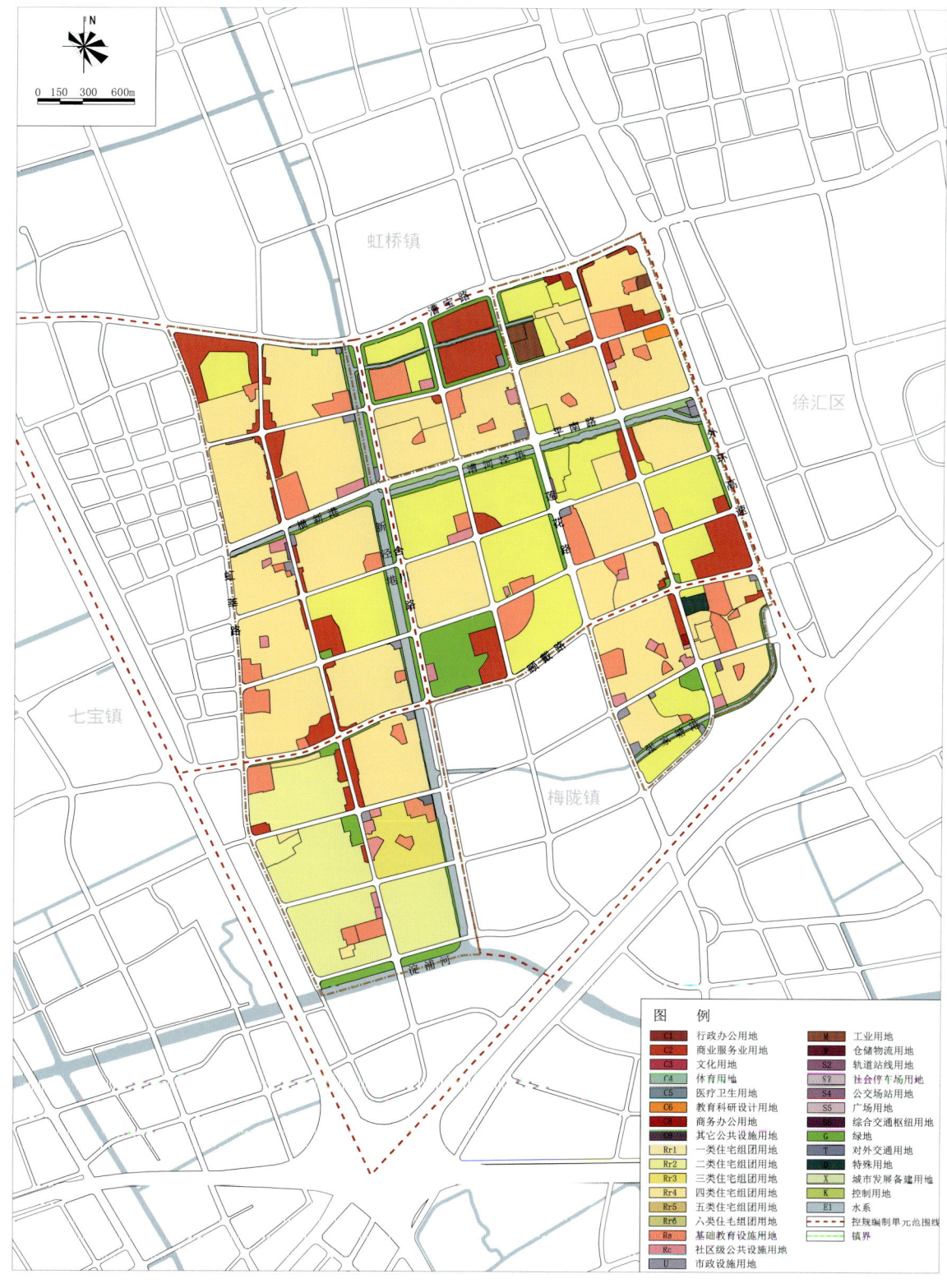

用地规划图（来源：古美路街道已批控制性详细规划拼合）
Land Use Planning (From: Control Detail Plan of Gumei Street)

第6章 地区风采 Part 6 Regional Elegance

万源城
Wanyuan City

古美艺术中心（来源：古美路街道办事处）
Gumei Art Center (From: Gumei Street Office)

十尚坊（来源：古美路街道办事处）
The Ten (From: Gumei Street Office)

第6章 地区风采 Part 6 Regional Elegance

梅陇镇
Meilong Town

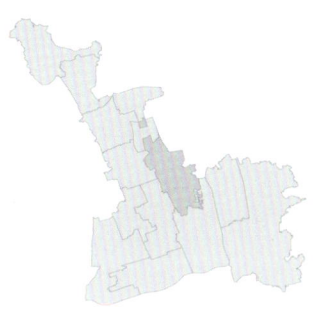

区位
Location

梅陇镇经济综合实力雄厚，是"全国 500 强乡镇"，辖区面积约 28.5 平方公里，常住人口 34.5 万人。梅陇，旧名梅家弄，得名于明朝梅姓徽商宅第似街弄。镇域内有百联南方购物中心、锦江乐园等商业办公和娱乐中心。随着梅陇地区中心、机场联络线华泾站周边地区、众欣产业园、欣梅产业园等地区城市更新的推进，梅陇将从城郊结合部逐步向"宜居、宜商、宜业"的现代化主城区迈进。

With a strong comprehensive economic strength, Meilong Town is listed in "China's Top 500 Villages and Towns" with an area of 28.5 square kilometers and a resident population of 345,000. Its former name Meijialong was named after the mansion of a Huizhou merchant in Ming Dynasty. Within the town, there are commercial offices and entertainment centers such as Bailian Nanfang Shopping Center and Jinjiang Park. Meilong will gradually change its function of suburban area, and become a "livable, business-friendly, and industry-friendly" modern central downtown, with the promotion of urban renewal in the center of Meilong, the surrounding area of Huajing Station on the terminal connecting line, Zhongxin Cultural Industrial Park, Xinmei Industrial Park, etc.

梅陇镇夜景
Night View of Meilong Town

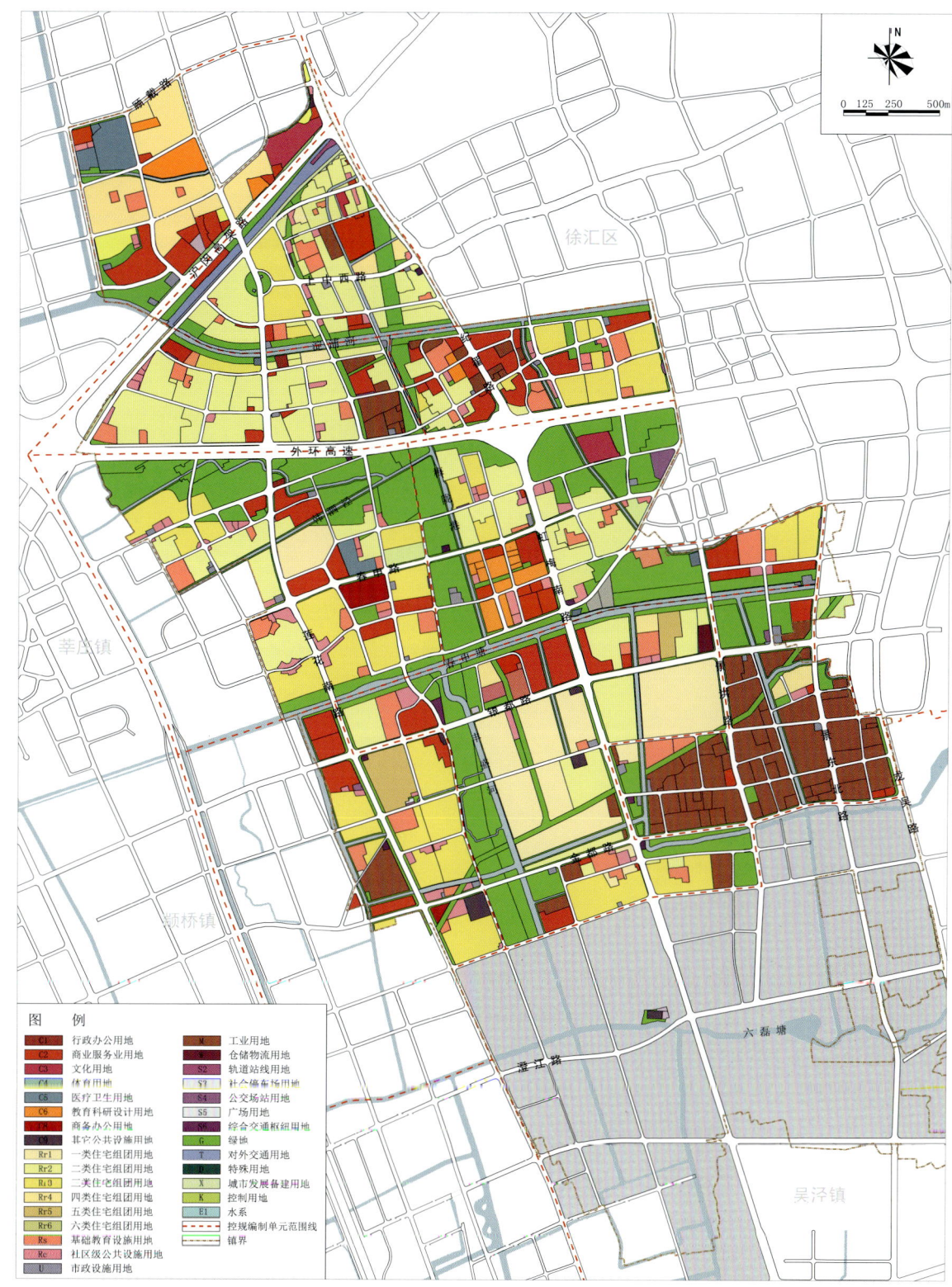

用地规划图（来源：梅陇镇已批控制性详细规划拼合）
Land Use Planning (From: Control Detail Plan of Meilong Town)

虹梅南路高架（来源：梅陇镇政府）
Hongmei South Elevated Road (From: Meilong Town Government)

城开中心
U Center City

美丽乡村永联村
Beautiful Yonglian Village

莘庄镇
Xinzhuang Town

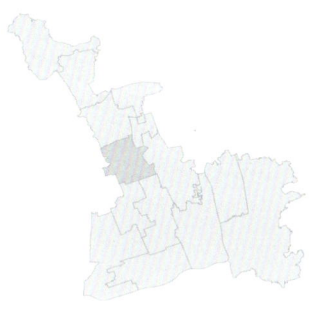

区位
Location

莘庄镇是闵行区区委、区政府所在地，是闵行区的政治、经济、文化中心，辖区面积约 19.1 平方公里，常住人口 28.1 万人。莘庄因跨莘溪而得名，历史悠久、交通便捷。自地铁 1 号线莘庄站建成运营，经过迅速发展，已经成为上海西南地区重要的新城。未来，结合主城副中心的定位提升，莘庄将承担区域综合服务职能，重点培育行政、文化、商业商务等功能，成为引领全区的功能核心。

Xinzhuang Town is the seat of Chinese Communist Party Committee of Minhang District and District Government, which makes it the political, economic, and cultural center of Minhang District. The town covers an area of 19.12 square kilometers and has a resident population of 281,000. Xinzhuang got its name because it crosses the Xinxi River, with a long history and convenient transportation. Since the establishment and operation of Xinzhuang Station of Metro Line 1, it has become an important new town in southwest Shanghai with the rapid development. In the future, in combination with the improvement of the positioning of the main city sub-center, Xinzhuang will assume the functions of regional comprehensive services by focusing on administrative, cultural, commercial functions, and become the core of the entire district.

莘庄镇
Xinzhuang Town

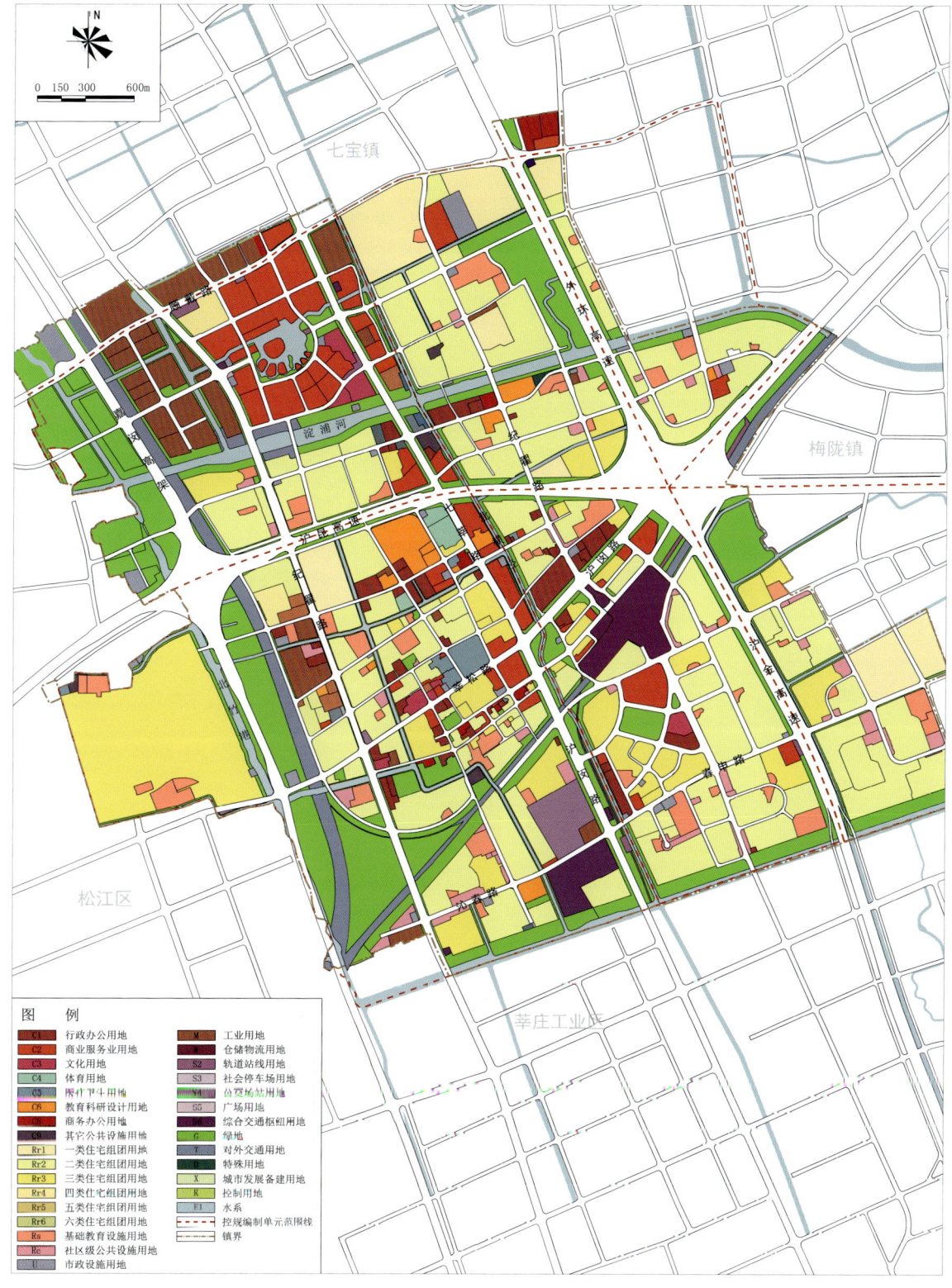

用地规划图（来源：莘庄镇已批控制性详细规划拼合）
Land Use Planning (From: Control Detail Plan of Xinzhuang Town)

第6章 地区风采 Part 6 Regional Elegance

莘庄立交
Xinzhuang Interchange

莘庄商务区
Xinzhuang Business District

莘庄镇居住区
Residential Area

第6章 地区风采 Part 6 Regional Elegance

165

颛桥镇
Zhuanqiao Town

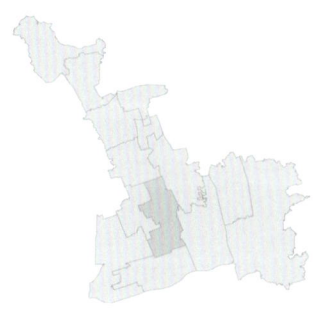

区位
Location

颛桥是"中国民间文化艺术之乡",辖区面积约32平方公里,常住人口27.4万人(含莘庄工业区)。始设于明代,拥有深厚的历史底蕴和丰富的人文风貌,"颛桥剪纸"已被列入上海市非物质文化遗产名录。作为闵行中部宜居宜业的地区代表之一,辖区内有向阳工业区、元江商务区以及规划元江地区中心,未来还将打造颛莘产城融合社区,营造面向创新人群、宜居宜业的城市环境。

Zhuanqiao is a town entitled "Chinese Folk Culture Art Village", with an area of 32 square kilometers and a resident population of 274,000 (including the Xinzhuang Industrial Zone). Founded in the Ming Dynasty, it has a long history and a rich culture. "Zhuanqiao Paper-cut" has been included in the Shanghai Intangible Cultural Heritage List. As one of the regional representatives of livable and suitable area for business in central Minhang, there are Xiangyang Industrial Zone, Yuanjiang Business Unit, and the planned Zhuanxin city-industry integration community will create a livable and suitable city environment for creative people.

颛桥镇
Zhuanqiao Town

用地规划图（来源：颛桥镇已批控制性详细规划拼合）
Land Use Planning (From: Control Detail Plan of Zhuanqiao Town)

第6章 地区风采 Part 6 Regional Elegance

光华路文创街区
Guanghua Road Cultural & Creative Block

剪纸公园
Paper-cut Park

颛桥历史文化长廊
Zhuanqiao History and Culture Corridor

第6章 地区风采 Part 6 Regional Elegance

莘庄工业区
Xinzhuang Industrial Zone

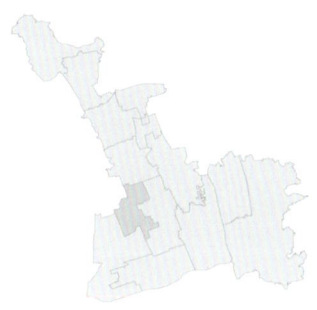

区位
Location

莘庄工业区是闵行重要的经济引擎。2018年，莘庄工业区实现工业总产值975.7亿元，财政总收入146.7亿元，是承载世界500强企业总部和研发总部、智能制造等企业的重要平台。未来，莘庄工业园将以产业研发和智能制造为主导，成为上海西南部产城融合产业新城。

Xinzhuang Industrial Zone is an important economic engine of Minhang. In 2018, the Xinzhuang Industrial Zone achieved a total industrial output value of 97.57 billion yuan and a total fiscal revenue of 14.67 billion yuan. It is an important platform for intelligent manufacturing, and the headquarters and R&D headquarters of firms of Fortune Global 500. In the future, Xinzhuang Industrial Zone will be led by industrial R&D and intelligent manufacturing, and will become a new city-industry integration zone in southwest Shanghai.

莘庄工业区（来源：莘庄工业区管委会）
Xinzhuang Industrial Zone (From:Xinzhuang Industrial Zone Management Committee)

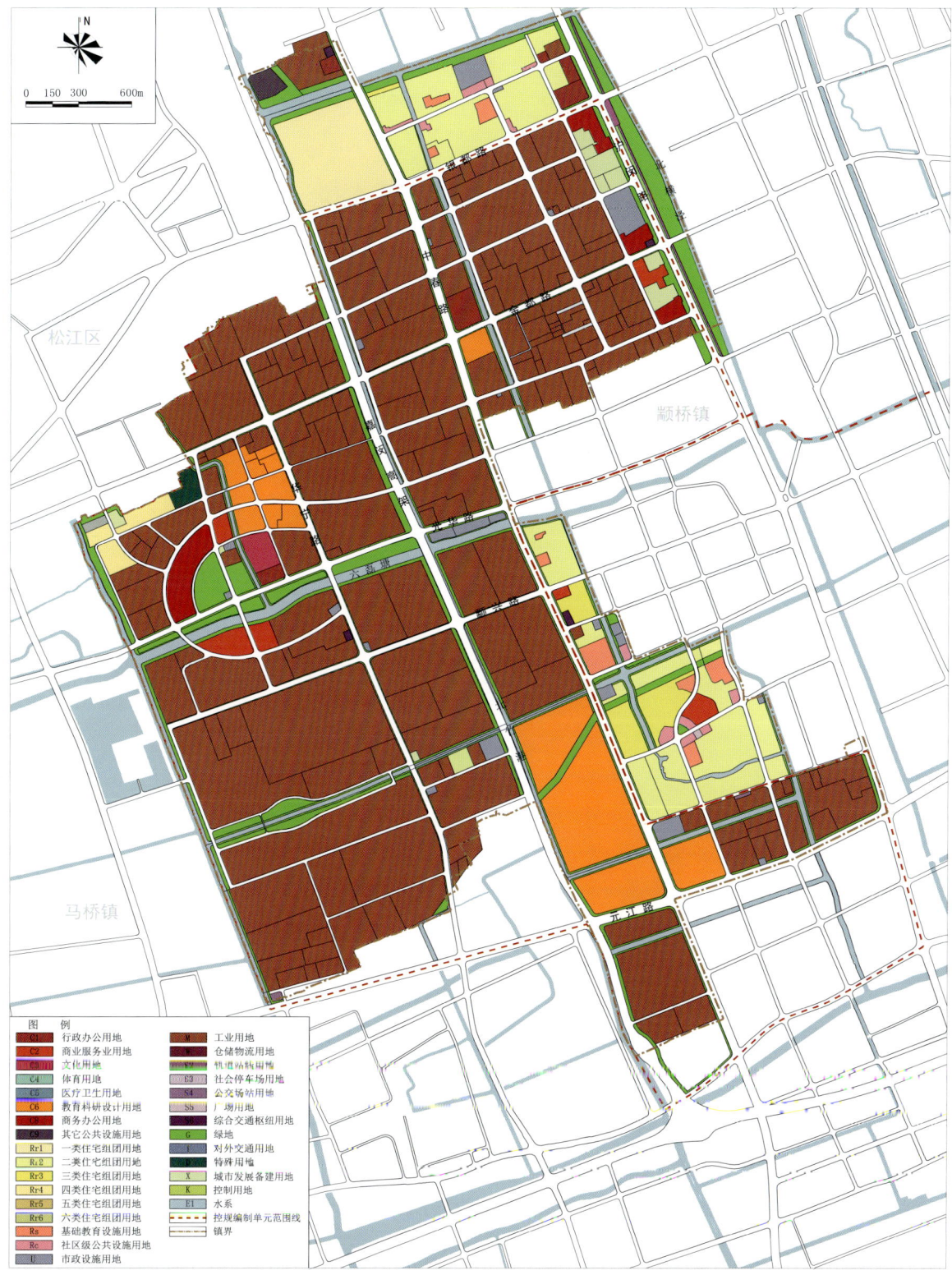

用地规划图（来源：莘庄工业区已批控制性详细规划拼合）
Land Use Planning (From: Control Detail Plan of Xinzhuang Industrial Zone)

第6章 地区风采 Part 6 Regional Elegance

上海航天技术研究院
Shanghai Academy of Spaceflight Technology

莘闵留创园（来源：莘庄工业区管委会）
Xinmin High-tech Park (From:Xinzhuang Industrial Zone Management Committee)

莘庄高新技术产业园
Xinzhuang High-tech Industrial Park

马桥镇
Maqiao Town

区位
Location

千年马桥，辖区面积49.5平方公里，常住人口11.2万人。拥有马桥古文化遗址、彭渡村传统村落、网球中心等资源，依托强大的文化魅力和网球大师赛等国际赛事，马桥已发展成为具备全球影响力的世界级体育赛事中心和国家级工业活力新城。未来将建设上海人工智能未来小镇，实现"生活更美好、生产更高效、生态更宜人"的发展目标。

Maqiao has an area of 49.5 square kilometers and a resident population of 112,000. With resources such as the Maqiao Ancient Culture Relic Site Park, the traditional village of Pengdu, and the Tennis Center, Maqiao has developed into a center that can hold world-class sports events and a town for national industrial zone. In the future, Maqiao will become a town of artificial intelligence in the future, and achieve the development goals of "better life, more efficient production, and more livable environment".

马桥镇
Maqiao Town

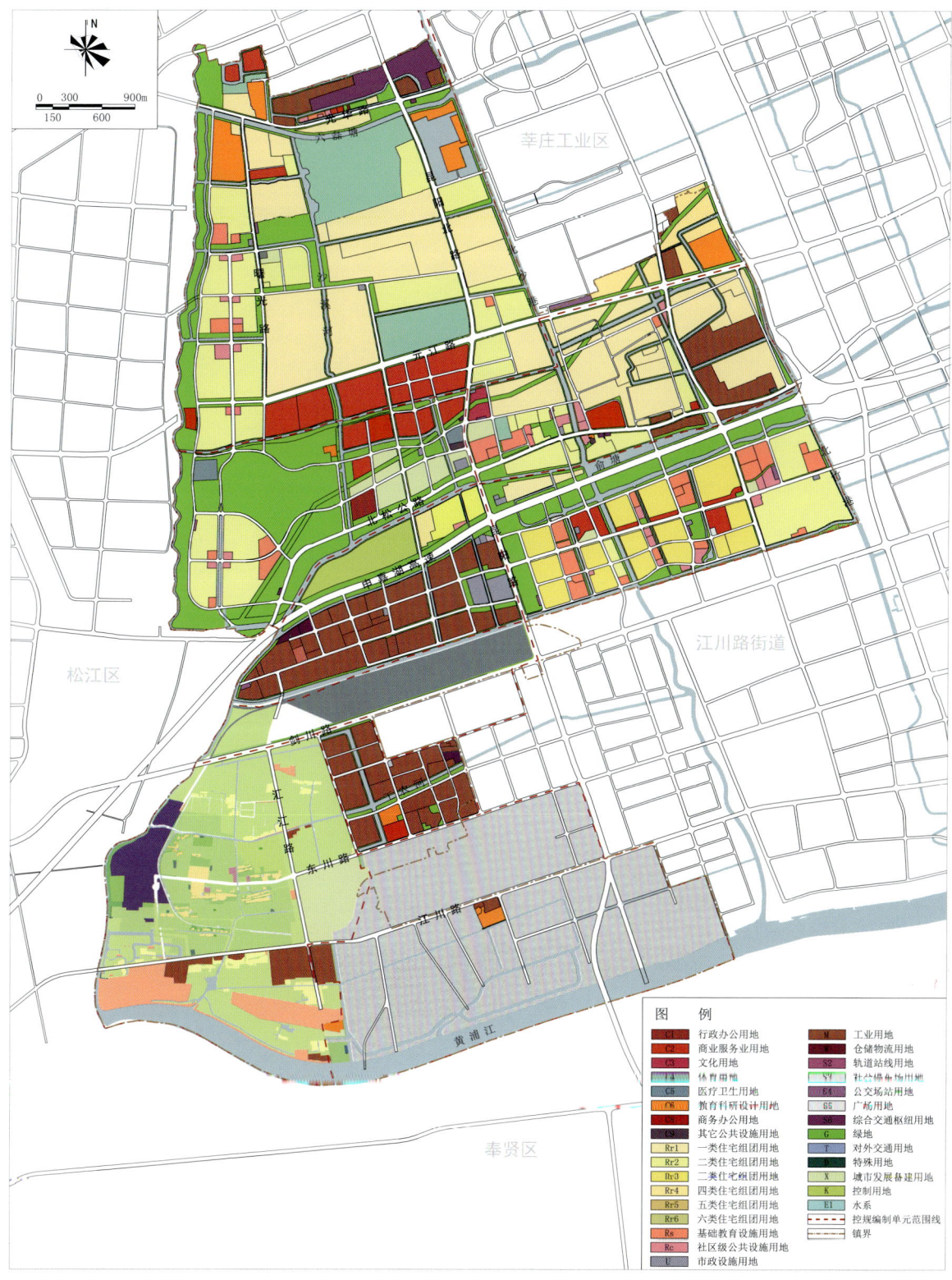

用地规划图（来源：马桥镇已批控制性详细规划拼合）
Land Use Planning (From: Control Detail Plan of Maqiao Town)

第6章 地区风采 Part 6 Regional Elegance

麦田（来源：马桥镇政府）
Wheat Field (From: Maqiao Town Government)

韩湘水博园（来源：马桥镇政府）
Hanxiang Water Expo Park (From: Maqiao Town Government)

旗忠森林体育城网球中心（来源：马桥镇政府）
Qizhong Forest Sport City Tennis Center (From: Maqiao Town Government)

江川路街道
Jiangchuan Street

区位
Location

历史上的"老闵行"地区，辖区面积约 27 平方公里，常住人口 19.3 万人。作为上海第一批卫星城镇，老闵行是重要的机电工业制造基地。闵行开发区及航空航天、高校所站推动了江川的改革创新。未来江川路街道将作为闵行区南部科技创新核心区的重要组成部分，积极推动城市更新，促进园区、校区、社区的三区联动，建设创新型产业社区。

The historical "old Minhang" area was about 27 square kilometers with a resident population of 193,000. As the first batch of satellite towns in Shanghai, the old Minhang is an important manufacturing base for the electromechanical industry. Minhang Development Zone, aerospace, and universities have promoted Jiangchuan's reform and innovation. In the future, Jiangchuan Street will be an important part of the core area of the technological innovation in the south of Minhang District. It will actively promote urban renewal, establish linkages between the park, campus, and community to build an innovative industrial community.

江川路街道（来源：江川路街道办事处）
Jiangchuan Street (From: Jiangchuan Street Office)

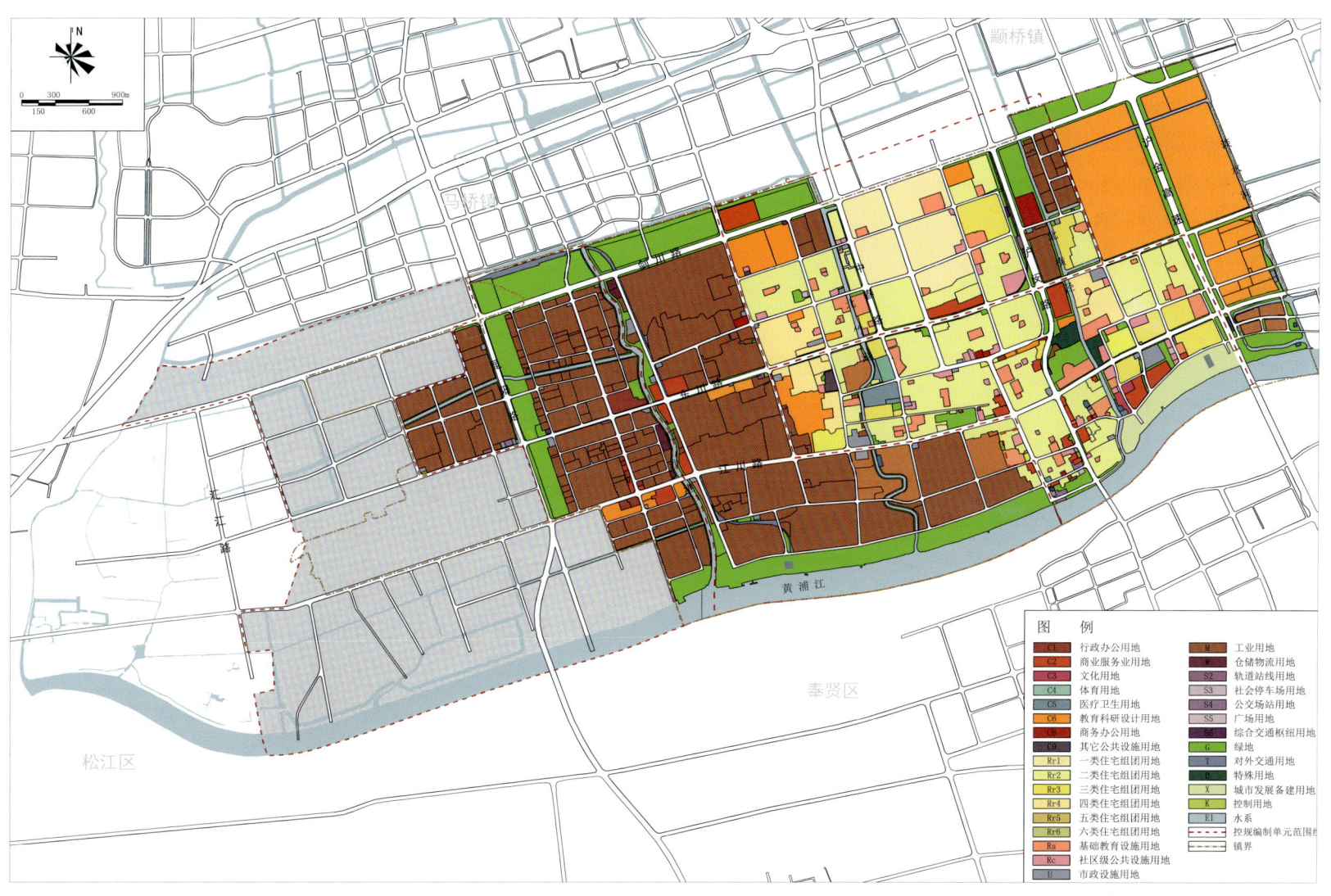

用地规划图（来源：江川路街道已批控制性详细规划拼合）
Land Use Planning (From: Control Detail Plan of Jianchuan Street)

第6章 地区风采　Part 6 Regional Elegance

江畔（来源：江川路街道办事处）
Huangpu Riverside (From: Jiangchuan Street Office)

江川路街道居住区
Residential Area

香樟1号路——江川路
Camphor No.1 Road— Jiangchuan Road

吴泾镇
Wujing Town

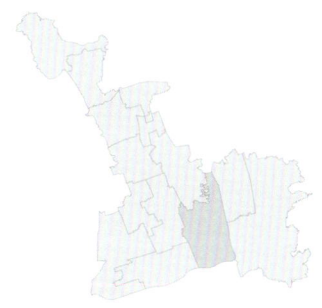

区位
Location

　　地处"浦江第一湾",辖区面积约 37.2 平方公里,常住人口 12.2 万人。得天独厚的岸线资源使得早期的吴泾成为上海西南水运主要物流周转地。龙吴路、剑川路、闵浦大桥带来的交通优势与紫竹高新区、上交大、华师大、航发商发等带来的产学研优势共同推动吴泾镇快速发展。未来吴泾将加快紫竹创新功能集聚区的建设,全力打造沪西南科创中心;加强滨江岸线、生态空间的塑造,完善生活、生态功能,彰显人文魅力。

　　Wujing Town is located in the "Pujiang First Bay" with an area of about 37.2 square kilometers and a resident population of 122,000. The unique shoreline resources made the early Wujing the main logistics turnover place of water transport in southwest Shanghai. The rapid development of Wujing is promoted by the transportation advantages brought by Longwu Road, Jianchuan Road, and Minpu Bridge, and the advantages of the integration of production, teaching and research brought by Zizhu High-tech Zone, Shanghai Jiao Tong University, East China Normal University, Aero Commercial Aircraft Engine Corporation, etc. In the future, Wujing will accelerate the construction of the industrial cluster with innovation functions in Zizhu, and make every effort to build the Technology Innovation Center in southwest of Shanghai, strengthen the shaping of the coastlines of Huangpu River and ecological space, improve the life and ecological functions, and highlight the cultural attraction.

吴泾镇
Wujing Town

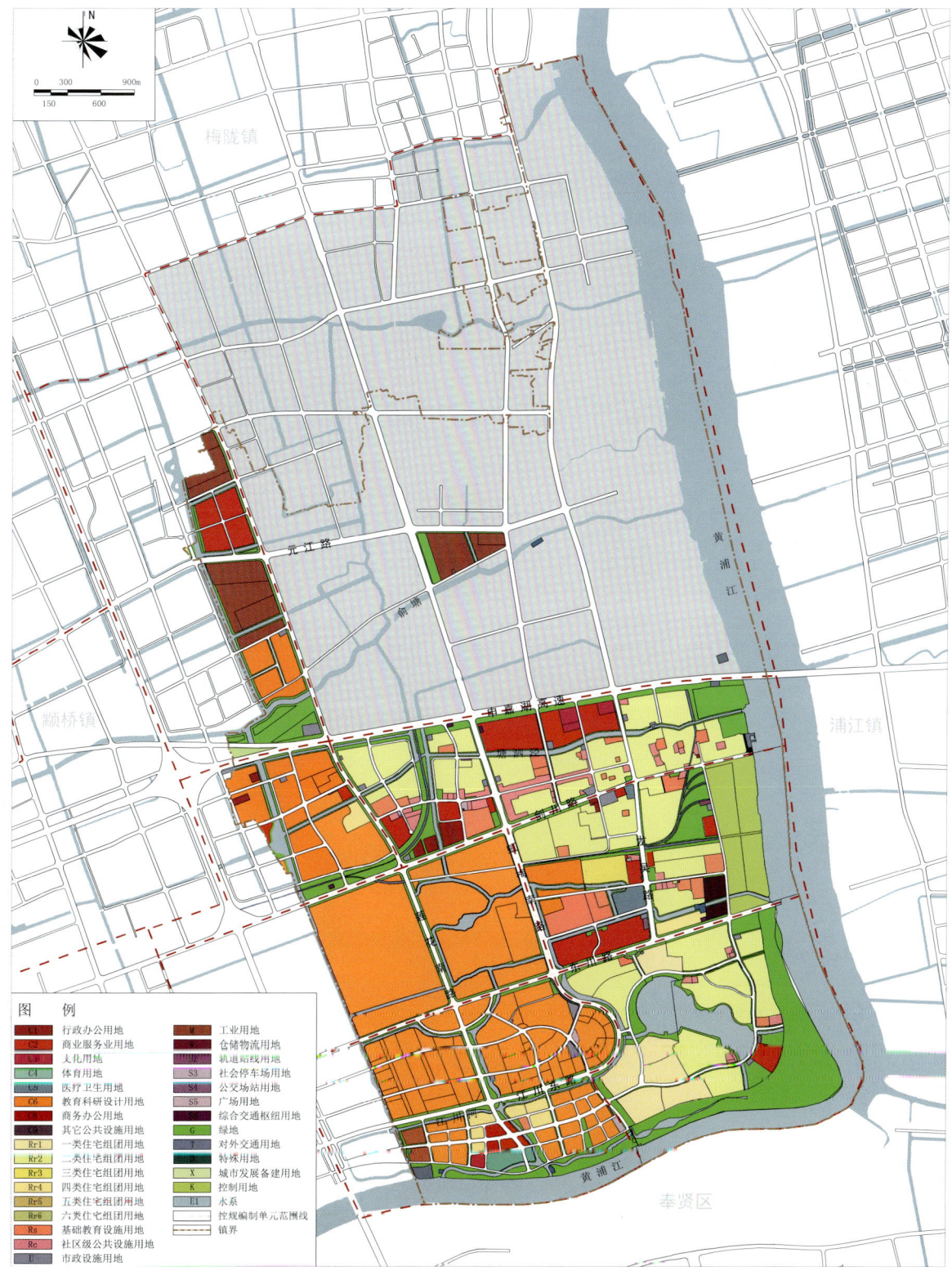

用地规划图（来源：吴泾镇已批控制性详细规划拼合）
Land Use Planning (From: Control Detail Plan of Wujing Town)

吴泾电厂
Wujing Power Plant

紫竹高新区
Zizhu High-tech Zone

184　　　心动上海，闵行正当年——规划建设二十年 THE RATE OF SHANGHAI　The Prime Time of Minhang

闵吴环卫综合码头
Minwu Sanitation Integrated Terminal

吴泾镇住区
Residential Area

浦锦街道
Pujin Street

区位
Location

　　浦锦街道是闵行最年轻的街道，成立于 2015 年，辖区面积约 24 平方公里。"浦"是代表新街道源自浦江镇的历史记忆，"锦"则蕴含对街道未来发展"锦上添花"、建设"锦绣家园"等美好祝愿和期待。拥有大片的生态林及涵养林，7000 亩（约 4.7 平方公里）连片的基本农田等生态资源，有华侨城、红醒半岛、保利茉莉等品质社区。未来浦锦街道将以花园社区为目标，进一步提升宜居水平和地区活力。

　　Pujin Street is the youngest street in Minhang. It was established in 2015 and covers an area of about 24 square kilometers. "Pu" represents the historical memory of the new street from Pujiang Town, and "Jin", which means "icing on the cake", contains beautiful wishes and expectations for the future development of the district, and the construction of a "splendid home". It has large-scale ecological forests and conservation forests, and 7,000 mu(4.7 square kilometers) of contiguous basic farmland and other ecological resources. There are quality communities such as Overseas Chinese Town, Hongxing Peninsula, Poly Moli Residence, etc. In the future, Pujin Street will take the garden community as the goal to further improve its livability and regional vitality.

浦锦街道（来源：浦锦街道办事处）
Pujin Street (From: Pujin Street Office)

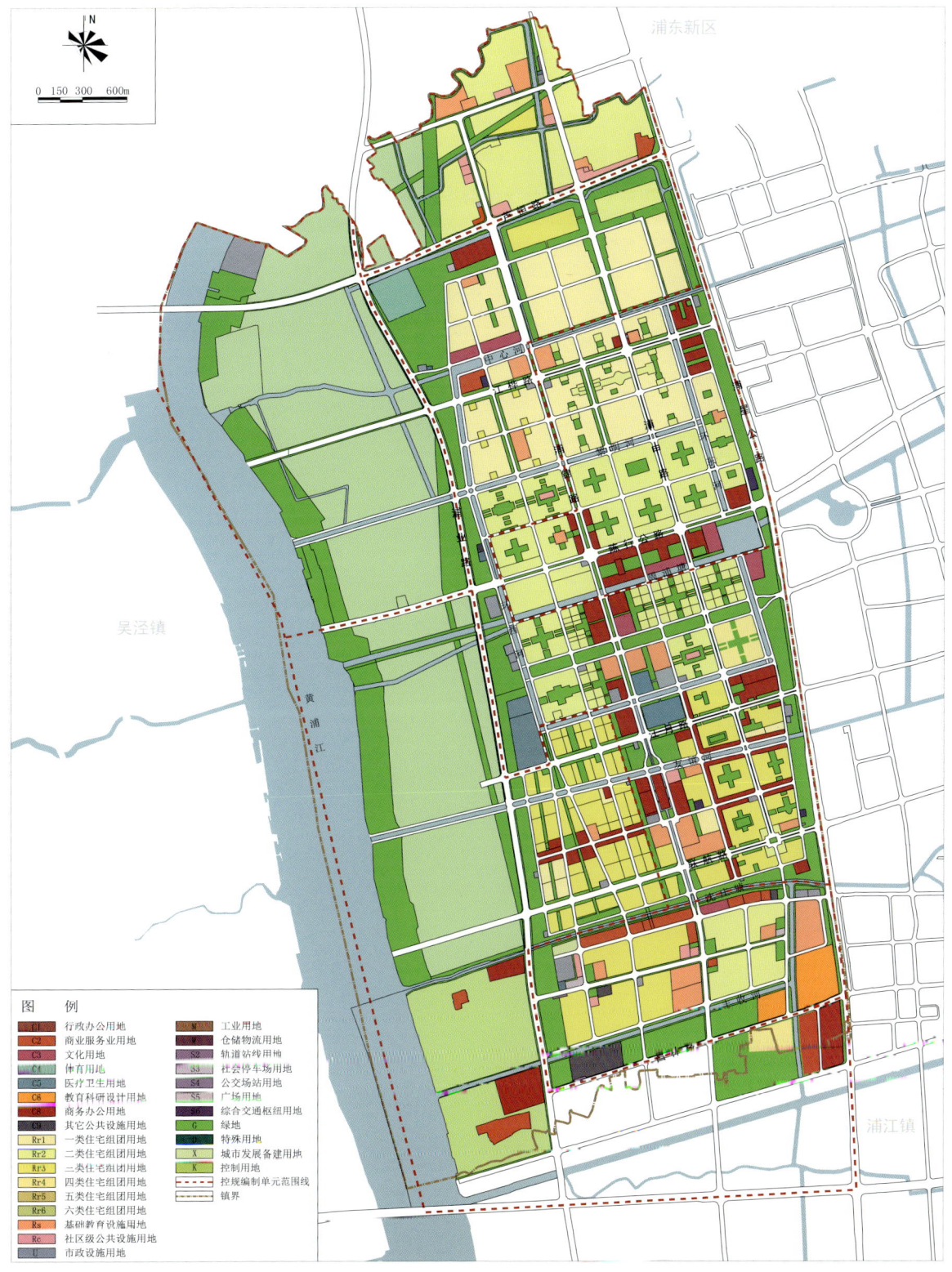

用地规划图（来源：浦锦街道已批控制性详细规划拼合）
Land Use Planning (From: Control Detail Plan of Pujin Street)

浦江幼儿园（来源：浦锦街道办事处）
Pujiang Kindergarten (From: Pujin Street Office)

浦锦绿地（来源：浦锦街道办事处）
Pujin Green Space (From: Pujin Street Office)

浦锦雕塑（来源：浦锦街道办事处）
Pujin Sculpture (From: Pujin Street Office)

第6章 地区风采 Part 6 Regional Elegance

浦江镇
Pujiang Town

区位
Location

浦江镇是多元融合的新市镇，辖区面积约78平方公里，常住人口约34.7万人（含浦锦街道）。拥有临港浦江园、航天产业园、闵东工业区等产业资源，召稼楼古镇、杜行老街等文化旅游资源，2万亩（约13.3平方公里）林地、4万亩（约26.7平方公里）基本农田、1000多条河道、浦江郊野公园等生态资源。未来，浦江镇将形成"一廊融产城、一脉串古今、片区提品质、郊野秀特色"的格局，建设生态、科技、人文的美丽新市镇。

Pujiang Town is a small and diverse town with an area of about 78 square kilometers and a resident population of about 347,000 (including Pujin Street). It has industrial resources such as Lingang Pujiang International Science & Technology City, Aerospace Park, Mindong Industrial Zone, etc., cultural tourism resources for example Zhaojialou Ancient Town, Duhang Old Street, etc., and ecological resources including 20,000 mu(13.3 square kilometers) of forest land, 40,000 mu(26.7 square kilometers) of basic farmland, more than 1,000 river channels, and Pujiang Countryside Park, etc. In the future, Pujiang Town will form an all-round industrial structure and build a beautiful new town of ecology, technology and humanities.

浦江镇
Pujiang Town

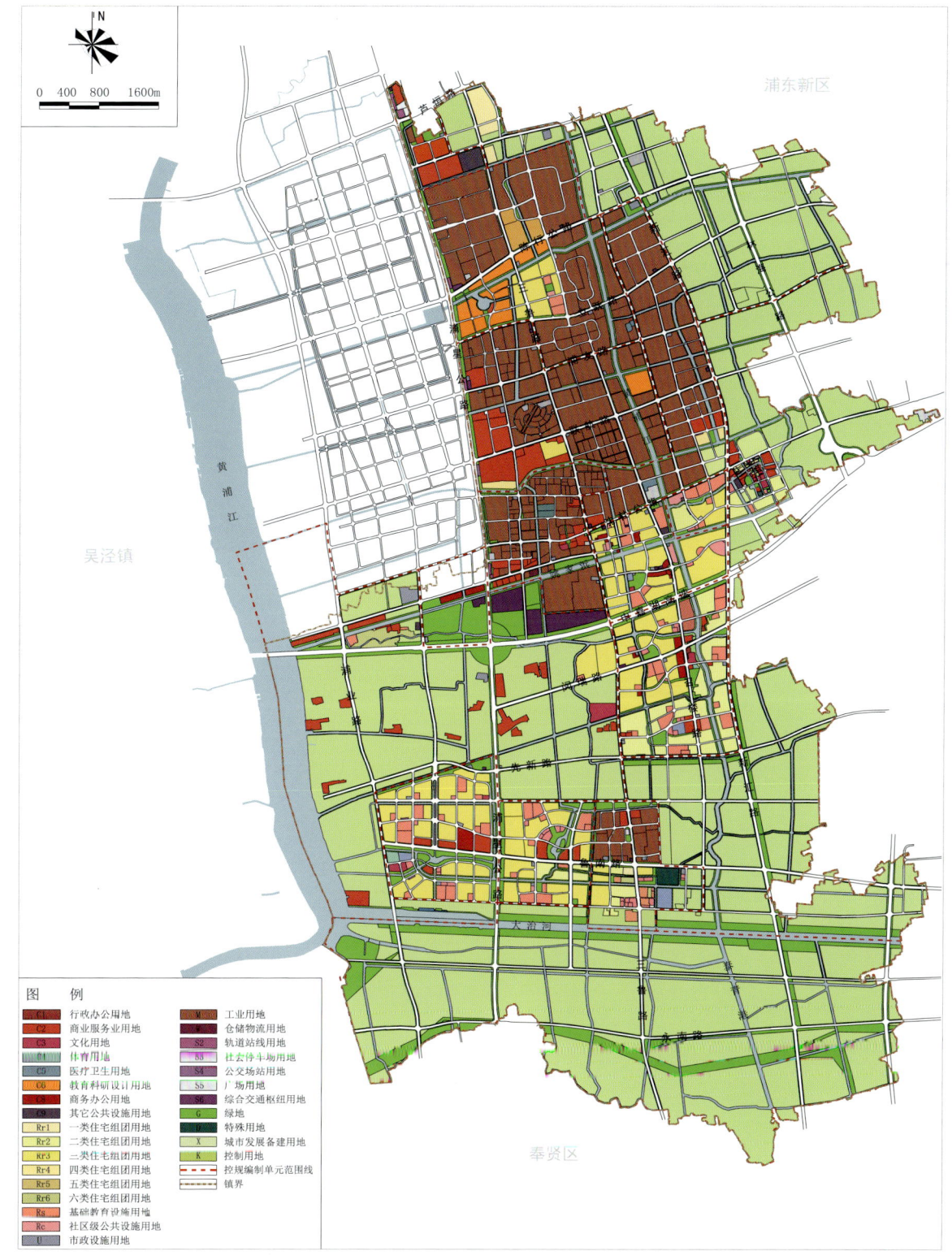

用地规划图（来源：浦江镇已批控制性详细规划拼合）
Land Use Planning (From: Control Detail Plan of Pujiang Town)

第6章 地区风采 Part 6 Regional Elegance

软件园
Science Park

革新村
Gexin Village

浦江新商圈（来源：浦江镇政府）
Pujiang New Business Zone (From:Pujiang Town Government)

第6章 地区风采 Part 6 Regional Elegance

后记
未来与挑战

随着上海 2035 总规批复、《长江三角洲区域一体化发展规划纲要》出台，闵行作为面向全球、面向未来，建设引领长三角地区更高质量一体化发展的国际开放枢纽的重要组成部分，将成为上海乃至长三角地区发展的强劲支点。虹桥国际开放枢纽、上海南部科创中心、莘庄副中心等重大板块，将助力闵行优化空间发展格局。未来闵行，充满了机遇。

但审视当下，前进的路程中也充满了挑战。

一、空间统筹发展难度大。传统以街镇和园区为主导的城市发展方式，使得空间破碎、功能混杂。未来较长一段时期内，亟需强化空间整合、优化功能布局、提升区域联动。以城市副中心和地区中心建设为抓手，承载落实市、区级的重要功能，实现空间与功能的整合。

二、紧约束条件下负重前行。闵行增量空间已经触顶，进入存量优化时代。未来，如何推进全区 75 平方公里低效用地转型和更新，减量现状建设用地 22 平方公里，撤并 90% 以上行政村将是急需攻克的难题。上海总规新增的两个城市副中心的规划建设、加快低效用地转型、寻找新的增长引擎、实现内涵式发展将是重中之重。

三、服务和品质提升千头万绪。公共服务体系虽已基本建成，但南北分布不均，公园覆盖率有待提高，品质与中心城区相比仍有一定差距。全面加快公共服务设施配置、完善公园体系建设、打造生态宜居闵行仍是关键任务。未来，全区将建成绿地 22.9 平方公里、绿道 400 公里，实现 3000 平方米以上公园、社区综合服务设施 500 米全覆盖，全面提高服务水平、提升环境品质、增加居民生活幸福感。

四、综合交通完善任重道远。全区交通矛盾诸多，整体路网密度偏低、道路联通性不足，路网格局有待完善；轨道交通覆盖率低，缺少局域线衔接，换乘不便。完善道路网络系统，加快轨道交通建设，优化公共交通体系刻不容缓。未来，全区路网密度将提升至 8 公里 / 平方公里，轨交站点 600 米覆盖率达 40%，交通出行快速便捷。

未来之路机遇与挑战并存，但闵行人执着守望、创新拼搏，相信未来必将成为创新者的乐土，奋斗者的乐园，繁荣繁华的主城区。

Postscript
The Future and Challenges

With the approval of "Comprehensive Plan of Shanghai (2017-2035)" and the issue of "The Integrated Regional Development of the Yangtze River Delta", Minhang, as a global and future-oriented, important component of building an international open hub that leads to higher-quality integrated development in the Yangtze River Delta region, will become strong fulcrum of development in Shanghai and even the Yangtze River Delta. Hongqiao International Open Hub, Shanghai South Science and Technology Innovation Center, Xinzhuang Sub-center and other major sectors will help Minhang optimize the spatial development pattern. The future of Minhang is full of opportunities.

Yet, the situations also pose great challenges to the development.

1. It is difficult to coordinate the development of space. The traditional urban development mode, which is dominated by streets and parks, makes the space fragmented and functions mixed. For a long period of time in the future, there is an urgent need to strengthen spatial integration, optimize functional layout, and enhance regional linkages. Taking the construction of urban sub-centers and regional centers as the starting point, it will carry out important functions at the city and district levels and realize the integration of space and functions.

2. It will carry forward with a heavy load under tight constraints. Minhang's incremental space has reached its peak and entered the era of inventory optimization. In the future, how to promote the transformation and renewal of 75 square kilometers of inefficient land in the region, reduce the current construction land by 22 square kilometers, and withdraw more than 90% of administrative villages will be an urgent problem to be overcome. The planning and construction of the two new urban sub-centers in the Shanghai Master Plan, accelerating the transformation of inefficient land, finding new growth engines, and achieving connotative development will be top priorities.

3. The service and quality improvement are complex. Although the public service system has been basically completed, it is unevenly distributed from north to south, the park coverage rate needs to be improved, and there is still a certain gap in quality compared to the central urban area. Accelerating the overall deployment of public service facilities, improving the construction of the park system, and creating an ecologically livable Minhang are still key tasks. In the future, the entire district will be completed with 22.9 square kilometers of green space and 400 kilometers of green roads. It will cover 500 meters of parks and comprehensive community service facilities with a total area of 3,000 square meters and above. Service levels, environmental quality, and happiness of residents will be improved comprehensively.

4. There is still a long way to go to improve comprehensive transportation. There are many traffic contradictions in the region. The overall density of the road network is low, the road connectivity is insufficient, and the road network needs further improvement. Also, the rail transit coverage rate is low, and transfer sites are insufficient, which makes it inconvenient for passengers to transfer. It is urgent to improve the road network system, speed up the construction of rail transit, and optimize the public transportation system. In the future, the density of the road network in the region will be increased to 8 km/km^2, and the coverage rate of rail crossing stations at 600 meters will reach 40%, making transportation fast and convenient.

Opportunities and challenges coexist on the road to the future, but Minhang people persist in watching and innovating, and believe that in the near future, Minhang will become a paradise for innovators, a paradise for strugglers, and a prosperous downtown.

编者按

本书立项于 2016 年底，正值我院紧密编制闵行 2035 总规之际，发现闵行需要一本能够囊括近 20 年发展建设成就的图书，来为闵行区的城市建设做一次总结。

四年间，从图书立项到策划编制，几易其稿，最终形成囊括闵行概览、产业发展、社会民生、生态休闲、乡村蜕变、地区风采六大章节的综合图书。成为我院为闵行区城市建设发展做出贡献的集成。

本书编制，是在区主要领导的关心下完成的。感谢闵行区规划和自然资源局、上海南虹桥投资开发有限公司、华漕镇、新虹街道、七宝镇、虹桥镇、古美路街道、梅陇镇、莘庄镇、莘庄工业区、颛桥镇、马桥镇、江川路街道、吴泾镇、浦锦街道、浦江镇等部门对本书编制提供的全力支持。

Editor's Notes

The book was established at the end of 2016, when Shanghai Minhang Planning Design and Research Institute was working closely on the Comprehensive Plan And General Land-use Plan of Minhang District ,Shanghai,2017-2035. We found that Minhang needed a book that could cover the development and construction achievements of nearly 20 years to make a summary of the urban construction of Minhang District.

In the past four years, it has undergone several revisions from establishment of the book to the planning and compilation.The book finally forms a comprehensive book covering six chapters: Minhang overview, industrial development, social livelihood, ecological leisure, rural metamorphosis and regional elegance. It has become the integration of our institute's contribution to the urban construction and development of Minhang District.

The compilation of the book was completed under the care of the main leaders of the district. Thanks to Minhang District Planning and Natural Resources Bureau, Shanghai Nanhongqiao Investment and Development Co., Ltd., Huacao Town, Xinhong Street, Qibao Town, Hongqiao Town, Gumei Street, Meilong Town, Xinzhuang Town, Xinzhuang Industrial Zone, Zhuanqiao Town, Maqiao Town, Jiangchuan Street, Wujing Town, Pujin Street, Pujiang Town and the other departments for their full support to the compilation of this book.

图书在版编目（CIP）数据

心动上海，闵行正当年：规划建设二十年：汉、英 / 上海闵行规划设计研究院有限公司编著. -- 上海：同济大学出版社，2020.12
ISBN 978-7-5608-9634-2

Ⅰ.①心… Ⅱ.①上… Ⅲ.①城市规划—研究—闵行—汉、英 Ⅳ.① TU984.251.3

中国版本图书馆 CIP 数据核字 (2020) 第 247883 号

心动上海，闵行正当年
规划建设二十年

上海闵行规划设计研究院有限公司　编著

策划编辑	江　岱	
责任编辑	周原田	
责任校对	徐春莲	
书籍设计	张　微	
出版发行	同济大学出版社　www.tongjipress.com.cn	
	（地址：上海市四平路 1239 号　邮编：200092　电话：021-65985622）	
经　销	全国各地新华书店	
印　刷	上海安枫印务有限公司	
开　本	787 mm×1092 mm　1/12	
印　张	17	
字　数	428 000	
版　次	2020 年 12 月第 1 版　2020 年 12 月第 1 次印刷	
书　号	ISBN 978-7-5608-9634-2	
定　价	128.00 元	

本书若有印装问题，请向本社发行部调换
版权所有　侵权必究